CONTENTS

FOREWORD

The request to write the Foreword to *Biochemistry for the Medical Sciences: An Integrated Case Approach* prompted me to reflect on the changes in the education and training of students in biochemistry during the last fifty years or so. In the 1940s and 1950s, it was possible for textbooks to cover the concepts and a majority of the facts of biochemistry; applications of the knowledge were limited and problem solving was relegated to the laboratory training of students. Three major changes have occurred in the intervening years: (1) the explosive increase in our biochemical knowledge; (2) the introduction of problem-solving approaches into mainstream teaching; and (3) the expectations and interest of students taking biochemistry courses.

As the amount of information expanded, textbooks of biochemistry became more encyclopaedic. Today authors must consciously limit the content, resorting more and more to multicoloured diagrams to express the complexity of topics. It is inconceivable for teachers to expect students to comprehend anything but a limited amount of the expansive database generated in the research laboratories around the world. Furthermore it is insufficient merely to present concepts and facts. Students want, and rightly expect, to see the relevance of biochemistry in a wide variety of areas, and this requires their interest to be stimulated by the presentation of biochemical knowledge in an everyday context. Just a few decades ago, most students in biochemistry were preparing for careers in medicine or biochemical research, but today individuals from a variety of disciplines are found in biochemistry courses. Students now want not just to acquire a basic biochemical knowledge but to recognize the relevance of biochemistry to their future careers, whether that be treating patients, or conducting research in biochemistry or in any other field.

This book by Steve Higgins, Tony Turner and Ed Wood adds a new, fascinating and essential resource for students and faculty. As stated in the Preface, this is not a textbook of biochemistry but rather a compilation of medically-related problems, ranging from famine in Africa to the adrenogenital syndrome, where a background in biochemistry is applicable. Although the authors assume the student has an elementary knowledge of biochemistry, the emphasis in this book is on the application of biochemical knowledge to problem solving.

The variety of problems presented demonstrates the breadth of the relationship of biochemistry to medicine. Cell biology, molecular biology, genetics, neurobiology, endocrinology and other fields are all featured, in addition to some metabolism. Medical problems as commonplace as cardiovascular disorders, anaemia, alcoholism and cancer are covered. The cases are not just statements of clinical conditions, but rather in each there is presented a practical and often personal approach. Presentation of clinical findings and normal population values places the problems into real situations for the reader. The structure of each problem permits a growth of knowledge. The *Questions* asked range from requests for basic facts to the integration of material. In all cases, the *Commentary* is sufficiently succinct to permit ready access to some of the facts necessary for understanding the problem, but is also expansive enough not to stifle or frustrate the student who is seeking a more in-depth discussion. Finally, the *Connections* are a unique and valuable resource for relating the problem to other biochemical areas and disciplines.

This book will be invaluable in a wide variety of teaching programs, but particularly in those programs involving a problem-based learning experience, whether in curricula largely founded on this type of teaching or in more traditional courses where problem-based learning exercises form an essential component alongside other teaching. These cases and the approach will stimulate extensive student discussions.

Thus, in conjunction with a basic textbook of biochemistry, this text should have a broad application in a variety of courses of biochemistry, not only in medical schools but also in undergraduate and graduate science courses. I am certain that students will find it valuable as they thread their way through the complexities of biochemistry.

Thomas M Devlin PhD
Philadelphia, Pennsylvania

May 1993

PREFACE

This is not a biochemistry text book. Rather, it is a collection of medically relevant problems that can be explained largely in biochemical terms. We hope that the problems will stimulate the interest of pre-clinical medical students, as well as other students of biochemistry, and lead to a fuller appreciation of the direct relevance of biochemistry and molecular biology to clinical medicine today.

Western society has seen dramatic decreases in childhood deaths through infectious diseases largely as a result of public health measures and mass immunisation programmes. Apart from accidental death, genetic disorders are now the most common paediatric caseload and are major contributors to adult morbidity and mortality. The application of the techniques of molecular and cell biology has led to major advances in diagnosis and treatment. Already we are beginning to gain an understanding of muscular dystrophy, Huntington's chorea, cystic fibrosis and even Alzheimer's disease, at the molecular genetic level. Genetic diseases that can be explained at the level of biochemistry have ceased to be confined to rarities that the clinician might encounter only very occasionally. A number of the problems in this book reflect these changes. Over the next 40 years (the clinical career of present-day graduates) familiarity with these techniques and their applications will be essential to appreciate fully the significance of these dramatic advances. Nevertheless, the average physician will continue to encounter cancers, atherosclerosis, anaemias, depression, etc. as a major part of the workload and, in the developing world, we will still have to deal with problems such as malnutrition, malaria, cholera and other parasitic infections.

The chapters in this book aim to reflect all these diverse aspects of modern medicine in a problem-solving format. They are mainly, but not exclusively, about diseases with a biochemical basis – some genetic, some rare and some more common. Some problems are short, others are longer and more demanding. They have all been chosen to cover some of the major areas of biochemistry, cell biology and molecular genetics that would, or should, be found in a modern, pre-clinical biochemistry course. In all cases the importance is the application of deductive reasoning, in line with current recommendations for medical education.

Each problem is structured in the same way and begins with a short *Introduction* which sets the scene. Next, the *Problem* itself is posed with the inclusion of such clinical and biochemical details as are necessary, using original data where possible. The cases are presented as accurately as possible, although we have taken the liberty of some simplifications so that basic principles are not obscured. To foster fluency in the language of medicine, the problems include many technical terms in common usage among doctors and clinical scientists; these are explained in the *Glossary* at the end of the book. Following the problem, there are a number of specific *Questions* which are structured so as to provide a stepwise guide towards a solution, usually a diagnosis of the disease and a possible management strategy. There follows a *Commentary*; we prefer not to use the term 'Answers' since, in many cases, alternative solutions may well be possible. *Further questions* relating to the problem and some up-to-date *References* are provided, normally to readily accessible reviews, for those who wish to study the topic further. Finally, in order to aid integration of the student's knowledge, *Connections* are made between the problem itself, other related problems in the book, and other relevant areas of biochemistry. The aim is to avoid biochemistry being seen as a set of discrete compartments (or lecture courses) which do not impinge on each other. For the more experienced, the *Connections* may seem obvious – but think about what additional connections *you* can provide. Our lists are by no means comprehensive! These various parts of the problem are arranged to be on different pages as far as possible; resist the temptation to 'peek' at the 'answers'.

We hope that we have been able to combine in this book the needs of both students and teachers. Pre-clinical medical students rightly demand medical relevance. Teachers want to emphasise principles not details, as well as the ability to reason and analyse data. All these components are available in this book. For teaching purposes the problems themselves can be used for small-group teaching or, alternatively, set for private study with the *Further Questions* being followed up in tutorial work. The Problems are presented roughly in the order of their difficulty but parts of a problem can be selected out to fit the stage which the students have reached in their course. We hope that the book will stimulate interest in biochemistry, emphasize its relevance to modern clinical practice and encourage lateral thinking. Happy problem-solving!

Steve Higgins
Tony Turner
Ed Wood
Leeds, 1993

ABBREVIATIONS

See also Appendix 5 for amino acid abbreviations.

Ab	antibody
ACE	angiotensin converting enzyme
ACTH	adrenocorticotropin
ADH	alcohol dehydrogenase
ADP	adenosine 5′-diphosphate
AIDS	acquired immuno-deficiency syndrome
ALA	δ-aminolævulinic acid
AldDH	aldehyde dehydrogenase
ALT	alanine transaminase
AMH	anti-Müllerian hormone
AMP	adenosine 5′-monophosphate
ANP	atrial natriuretic peptide
ASO	allele-specific oligonucleotide
AST	aspartate transaminase
ATP	adenosine 5′-triphosphate
cAMP	adenosine 3′,5′-cyclic monophosphate
CAT	chloramphenicol acetyl-transferase
cDNA	complementary DNA
cGMP	guanosine 3′,5′-cyclic monophosphate
CK	creatine kinase
CoA	coenzyme A
Con A	concanavalin A
CoQ	coenzyme Q (ubiquinone)
DHPR	dihydropteridine reductase
DHT	5α-dihydrotestosterone
DIPF	diisopropylphosphofluoridate
DNA	deoxyribonucleic acid
DNOC	*ortho*-dinitrocresol
2-DOG	2-deoxyglucose
DOPA	dihydroxyphenylalanine
EDTA	ethylene diamine tetraacetic acid
ELISA	enzyme-linked immunosorbent assay
ER	endoplasmic reticulum
$FADH_2$	flavin adenine dinucleotide (reduced form)
FH	familial hypercholesterolaemia
GABA	γ-aminobutyric acid
γ-GT	γ-glutamyl transferase
GMP	guanosine 5′-monophosphate
GOT	glutamate-oxaloacetate transaminase
G protein	guanosine nucleotide-binding protein; G_s, G_i, G_q, G_t, are G protein subtypes
GPT	glutamate-pyruvate transaminase
GTP	guanosine 5′-triphosphate
GTPase	guanosine 5′-triphosphatase
Hb	haemoglobin
Hb CS	haemoglobin variant Constant Spring
hCG	human chorionic gonadotropin
HDL	high-density lipoprotein

HIV — human immunodeficiency virus
HLA — human leukocyte-associated antigens
HMG-CoA — 3-hydroxy-3-methyl glutaryl-CoA
HPLC — high performance liquid chromatography
5-HT — 5-hydroxytryptamine

IQ — intelligence quotient

kb — kilo base pairs
K_d — dissociation constant
kDa — kilo Dalton
K_m — Michaelis constant

LDL — low-density lipoprotein
LH — luteinising hormone
LT — labile enterotoxin

MAO — monoamine oxidase
MCH — mean corpuscular haemoglobin
MCHC — mean corpuscular haemoglobin concentration
MCV — mean corpuscular volume
MHC — major histocompatibility complex
MPS — mucopolysaccharide
MPTP — 1-methyl-4-phenyl-1,2,5,6-tetrahydropyridine
M_r — relative molecular mass
mRNA — messenger RNA
MSA — multiple system atrophy

NAD^+ (NADH) — nicotinamide adenine dinucleotide (reduced form)
$NADP^+$ (NADPH) — nicotinamide adenine dinucleotide phosphate (reduced form)
NHNGP — *N*-hydroxy-2-naphthylamine-β-1-glucuronyl pyranoside
NT — neurotransmitter

PAF — pure autonomic failure
PCR — polymerase chain reaction
PCV — packed cell volume
PD — Parkinson's disease
PET — positron emission tomography
PFK — phosphofructokinase
PKU — phenylketonuria
PTH — parathyroid hormone

RIA — radioimmunoassay
RNA — ribonucleic acid

SD — standard deviation
SDS — sodium dodecylsulphate
SDS–PAGE — polyacrylamide gel electrophoresis in the presence of SDS

T_3 — triiodothyronine
T_4 — thyroxine
TDF — testis determining factor
TIBC — total iron binding capacity
tPA — tissue-type plasminogen activator
TRH — thyrotropin releasing hormone
TSH — thyroid-stimulating hormone (thyrotropin)

UDP — uridine 5′-diphosphate

v — initial rate (of an enzyme reaction)
VLDL — very low-density lipoprotein
V_{max} — maximum rate (of an enzyme reaction)

ACKNOWLEDGEMENTS

We are grateful to our colleagues, Alison Caswell, John Illingworth and Tom Scott, who suggested some of the problems included in this book. Many of the problems have been used, albeit in embryonic form, as a highly successful part of the Preclinical Biochemistry course in the University of Leeds Medical School. We thank all of our many colleagues and students who have given us their opinions on these deductive exercises over the years since their inception.

We place great emphasis on using original data in the problems. Many of our research colleagues from around the world have generously provided photographs and other illustrative material for the book. We are profoundly grateful to them and acknowledge their contributions in the figure legends.

We also thank the following who gave us their invaluable advice on the biochemical and clinical aspects of the cases and who commented on the text:

Herman Bachelard	Department of Physics, University of Nottingham
Julian Barth	Department of Chemical Pathology, University of Leeds
David Bender	Department of Biochemistry & Molecular Biology, University College, London
John Clegg	Institute of Molecular Medicine, John Radcliffe Hospital, University of Oxford
Steve Goodall	Department of Chemical Pathology, University of Leeds
Mike Grant	Department of Biochemistry & Molecular Biology, University of Manchester
Simon van Heyningen	Department of Biochemistry, University of Edinburgh
Francois van Hoof	International Institute of Cellular & Molecular Pathology, Université Catholique de Louvain, Brussels, Belgium
Nigel Hooper	Department of Biochemistry & Molecular Biology, University of Leeds
Miles Houslay	Department of Biochemistry, University of Glasgow
Louis Hue	International Institute of Cellular & Molecular Pathology, Université Catholique de Louvain, Brussels, Belgium
Paul Luzio	Department of Clinical Biochemistry, Addenbrooke's Hospital, University of Cambridge

Howard Parish	Department of Biochemistry & Molecular Biology, University of Leeds
Malcolm Parker	Imperial Cancer Research Fund Laboratories, Lincoln's Inn Fields, London
David Potts	Department of Physiology, University of Leeds
Guy Rousseau	International Institute of Cellular & Molecular Pathology, Université Catholique de Louvain, Brussels, Belgium
Margaret dos Santos Medeiros	Department of Biochemistry & Molecular Biology, University of Leeds
Ken Siddle	Department of Clinical Biochemistry, Addenbrooke's Hospital, University of Cambridge
Anne Soutar	Medical Research Council Lipoprotein Team, Hammersmith Hospital, London
Don Steiner	Howard Hughes Medical Research Institute, Department of Biochemistry & Molecular Biology, University of Chicago, Illinois, USA
Rod Thompson	Department of Clinical Biochemistry, University of Southampton
Keith Tipton	Department of Biochemistry, Trinity College Dublin, Ireland
Colin Toothill	Department of Chemical Pathology, University of Leeds
Frank Vella	Department of Biochemistry, University of Saskatchewan, Canada
Gillian Wallis	Department of Biochemistry & Molecular Biology, University of Manchester
Malcolm Watford	Department of Nutritional Sciences, Rutgers State University, New Jersey, USA
David Weatherall	Institute of Molecular Medicine, John Radcliffe Hospital, University of Oxford
Judy Wyatt	Department of Histopathology, St James's University Hospital, Leeds

We are grateful to the following for permission to reproduce copyright material:

The Catholic Fund for Overseas Development for Fig. 2.1; The World Health Organization for Fig. 2.2; The New England Journal of Medicine and the author, Prof. B. Sobel, for Fig. 3.3 (Van de Werf *et al.* 1984); Plenum Publishing Corp. and the authors, Drs M. I. New and L. S. Levine, for Figs 7.2 and 7.5 (Harris and Kirschorn (eds) 1973); The American Association for Biochemistry and Molecular Biology and the author, Prof. D. J. Prockop, for Figs 8.3 and 8.4 (Vogel *et al.* 1988); Pergamon Press Ltd. and the author, Prof. D. J. Prockop, for Fig. 8.5 (Prockop *et al.* 1989); Prof. W. B. Greenough, Dr. N. Hirschhorn and Dr. M. Robbins for Fig. 13.1

(Hirschhorn and Greenough 1991); Rhone-Poulenc Rorer Pharmaceuticals for Fig. 13.2; Blackwell Scientific Publications for Fig. 13.3 (Finean *et al.* 1979); The British Medical Journal for Fig. 17.1 (Farrell and Strang 1992); The British Medical Journal and the author, Dr. G. R. Thompson, for Fig. 20.6 (Barbir *et al.* 1989); Macmillan Magazines Ltd for Figs 21.3 and 21.6.

Whilst every effort has been made to trace the owners of copyright material, in a few cases this has proved impossible and we take this opportunity to offer our apologies to any copyright holders whose rights we may have unwittingly infringed.

CONCENTRATION TERMS

We have used the biochemical convention such that a concentration of 10 mmol/l is written as 10 mM. In the USA, non-systematic concentration terms are still commonly used; for instance 10 mM glucose is often expressed as 180 mg/dl.

Make sure you know the distinction between an **amount** (mol) and a **concentration** (mol/l or M). The concentrations and amounts of many compounds in the body (e.g. vitamins, hormones) are exceedingly small so you will come across the following prefixes:

d (deci) $= 10^{-1}$	n (nano) $= 10^{-9}$
m (milli) $= 10^{-3}$	p (pico) $= 10^{-12}$
μ (micro) $= 10^{-6}$	f (femto) $= 10^{-15}$

1 TEENAGE WEAKLING

Introduction

Skeletal muscle has a highly variable requirement for energy (in the form of ATP) depending on the demand placed on the muscle. Potentially, this energy can be derived from a variety of sources. However, during severe exercise when the energy demand is very high, blood is squeezed out of the muscle bed by contractile activity. The muscle is then temporarily starved of oxygen (becomes ischaemic), making oxidative processes very difficult. In these circumstances, oxidation of fatty acids, normally the most efficient means of generating energy, cannot be relied upon to satisfy the large energy demand. Instead the muscle has to depend more heavily on anaerobic glycolysis. Even here the muscle faces problems since the diminished blood flow prevents glucose being imported to fuel glycolysis so the muscle's glycogen stores must be mobilised. In addition, the end product of anaerobic glycolysis, lactate, cannot be removed efficiently via the blood and has to accumulate within the muscle. As a consequence, humans cannot carry out severe ischaemic exercise for more than a few minutes, as athletes will know only too well (Figure 1.1).

Comparatively rarely, individuals are encountered who, because of a genetic defect, cannot tolerate ischaemic exercise. This problem features such an individual.

Figure 1.1. Sprinter crossing finishing line.

The Problem

Phillip M., a sixteen-year-old boy, sought medical help for progressive muscle weakness over a number of years. He experienced painful muscle cramps on severe exercise but he could tolerate moderate exercise normally. The effect of severe ischaemic exercise on blood lactate is shown in Figure 1.2. Severe exercise was followed by dramatically elevated serum levels of lactate dehydrogenase, creatine kinase and aldolase, which persisted for some hours after exercise had ceased. Myoglobinuria was also present and there was evidence of mild haemolysis. Phillip's blood glucose level was normal and could be elevated by treatment with glucagon. It was decided to do a muscle biopsy. Table 1.1 shows the results of the biopsy analysis.

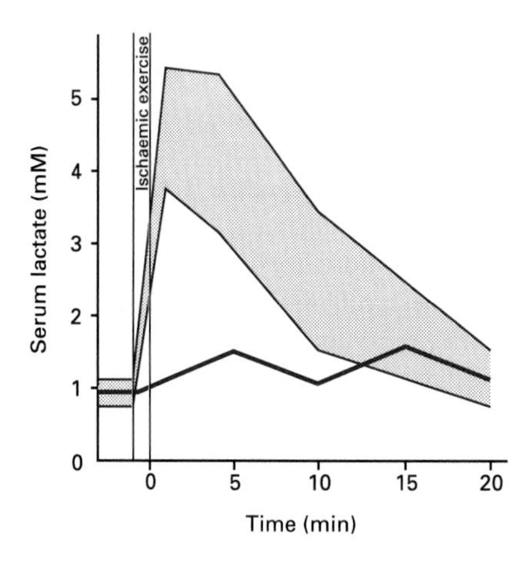

Figure 1.2. Effect of ischaemic work of the forearm muscles on blood lactate. Data for the patient are shown (solid line) with results for normal subjects (shaded area).

Table 1.1. Analysis of metabolites in muscle biopsy samples.

Source of muscle	Tissue component: Glycogen (mg/g tissue)	Glucose 6-phosphate (μmol/g tissue)	Fructose 6-phosphate (μmol/g tissue)	Fructose 1,6-bisphosphate (μmol/g tissue)
Phillip	43.8	9.2	1.6	0.02
Normal volunteers (mean ± SD)	9.6 ± 1.8	0.5 ± 0.3	0.1 ± 0.05	0.61 ± 0.23

Questions

1. Why is lactate, rather than pyruvate, produced by normal muscle when it is working anaerobically?
2. What do the results of the ischaemic exercise test suggest is wrong with Phillip's muscles?
3. How does glucagon act and what do the results of treating Phillip with glucagon indicate about his condition?
4. What can you conclude from the muscle biopsy analysis?
5. From your deductions so far, what do you think is the metabolic defect in the patient?
6. Why is Phillip able to tolerate moderate, but not severe, exercise?
7. What is the significance of his serum enzyme levels and the myoglobinuria?
8. Why is there an increased deposition of muscle glycogen (Table 1.1)?
9. What evidence might have been present to support the suggestion of haemolysis? How does it arise in this patient?

Commentary

During ischaemic exercise, skeletal muscle normally derives its energy via glycolysis utilising glucose 6-phosphate mobilised from muscle glycogen stores. To regenerate NAD^+ from the NADH produced in glycolysis, and so sustain glycolytic flux, pyruvate must be reduced to lactate. As a result, the muscles produce large amounts of lactate which eventually enters the blood and is later taken up by the liver to be reformed into glucose via gluconeogenesis (the Cori cycle). The ischaemic exercise test suggested that Phillip cannot sustain a significant rate of glycolysis. Examination of the levels of glycolytic intermediates in the muscle biopsy suggests that the patient has a deficiency in the muscle glycolytic enzyme, phosphofructokinase (PFK), as indicated by the elevated levels of glucose 6-phosphate and fructose 6-phosphate coupled with decreased fructose 1,6-*bis*phosphate, the product of the PFK-catalysed reaction.

Glucagon is a polypeptide hormone secreted by the α cells of the islets of Langerhans in the pancreas in response to lowered blood glucose concentration. It acts on the liver via a cascade mechanism involving cyclic AMP and protein kinase to stimulate the mobilisation of hepatic glycogen and replenishment of blood glucose. Thus the glucagon test indicates that an inability to maintain his blood glucose concentration is not the explanation for Phillip's defective muscle glycolysis. However, even though his glycolytic pathway is inefficient, and energy production cannot be sustained during ischaemic exercise, Phillip tolerated moderate exercise for two reasons. Firstly, PFK is not completely absent (it would probably have been fatal *in utero*) since some glycolytic flux does occur. Secondly, in moderate exercise the muscles would be working largely aerobically using fatty acids as fuels via β-oxidation and the tricarboxylic acid cycle.

Although the defect does not involve an enzyme directly concerned with glycogen synthesis, increased deposition of glycogen occurs (Table 1.1). Indeed this feature of the condition prompted the discoverer of the disease, Seiichiro Tarui, to describe it as a glycogen storage disorder (glycogenosis type VII). Since glycolysis is partially blocked, glucose 6-phosphate accumulates (see Table 1.1). Although one might be tempted to explain the increased synthesis of glycogen simply in terms of a mass action effect, glucose 6-phosphate is actually an allosteric activator of one form of glycogen synthase (the relatively inactive or phosphorylated form).

A large part of the energy expended by a cell is required to maintain its membrane integrity and polarity via various active transport systems. During ischaemic exercise Phillip's muscle cells are energy deficient. Consequently, membrane integrity is compromised and soluble proteins (such as lactate dehydrogenase, creatine kinase, aldolase and myoglobin) leak out from the muscle cell into the extracellular fluid. The proteins investigated in this case are not the only ones that would be present in Phillip's serum, but are some of the most diagnostic and readily assayed.

There are two PFK isoenzymes, the muscle and liver forms. Phillip's defect lies in the gene for the muscle isoenzyme and thus his liver metabolism is unaffected. Erythrocytes are totally dependent on glycolysis for their energy and are particularly vulnerable to glycolytic disorders. So it is to be expected that Phillip's condition will involve impaired erythrocyte viability, with consequent haemolysis and elevated serum bilirubin. However, because erythrocytes express both the muscle and liver isoenzymes of PFK, glycolysis is only partially compromised in Phillip's erythrocytes.

Unfortunately, Phillip cannot be offered a cure, but provided he avoids strenuous exercise he can expect to lead a relatively normal life.

Further Questions

1. The glycogen deposited in increased amounts in Phillip's muscle had a normal structure. What does 'normal' glycogen mean to you? In what ways is the structure of glycogen 'abnormal' in some other glycogen storage disorders?
2. Both cardiac muscle and skeletal muscle working aerobically utilise fatty acids as fuels in preference to glucose or glycogen. What are the control mechanisms involved in this choice?
3. How could blood lactate be assayed? What are the essential requirements for a valid assay?
4. Glucose 6-phosphate can be assayed spectrophotometrically using glucose 6-phosphate dehydrogenase and $NADP^+$. How would you modify the assay to allow you to quantify fructose 6-phosphate and fructose 1,6-*bis*phosphate in a muscle extract?
5. List the different types of muscle fibre in skeletal muscle. What type would be affected in Phillip's condition?
6. PFK catalyses the formation of fructose 1,6-*bis*phosphate. Fructose 2,6-*bis*phosphate also occurs in muscle. What is its role?

Connections

- Draw outline schemes for the processes of glycolysis, gluconeogenesis, glycogenesis (glycogen synthesis) and glycogenolysis (glycogen breakdown). Fit these processes together and explain the mechanisms which ensure that the synthetic and degradative pathways do not operate simultaneously.
- Review the glycogen storage disorders, their biochemistry, the medical symptoms and the prognoses.
- List the different processes whereby ATP may be regenerated by phosphorylation of ADP. What type of phosphorylation is involved in the production of ATP in anaerobic glycolysis and at which points in the pathway is ATP produced? What is the significance of creatine kinase in muscle? (See also Problems 3 and 4.)
- Use this problem to review your knowledge of how insulin and glucagon influence carbohydrate metabolism and the different roles played by liver and muscle glycogen. (See also Problem 12.)
- Check that you understand what jaundice is and the different ways in which it may arise. (See also Problems 6 and 11.)
- Think of ways in which enzymes can be used as diagnostic tools or therapeutic reagents. (See also Problem 3.)

References

Danon M J, Serviei S, DiMauro S and Vora S (1988) Late-onset muscle phosphofructokinase deficiency. *Neurology* **38**, 956–60.

Stanby W C and Connett R J (1991) Regulation of muscle carbohydrate metabolism during exercise. *FASEB Journal* **5**, 2155–9.

Tanaka K R and Zerez C R (1990) Red cell enzymopathies of the glycolytic pathway. *Seminars in Hematology* **27**, 165–85.

2 FAMINE IN AFRICA

Introduction

We have all witnessed through the medium of television and the newspapers harrowing pictures of starving children dying in Africa, such as that shown in Figure 2.1. Our immediate reactions are of compassion and we feel compelled by our common humanity to contribute what we can to charitable agencies engaged in famine relief. Almost certainly we will also feel anger and helplessness in the face of the apparently insuperable problems of drought, wars, foreign debt, local corruption and adverse trading arrangements with Western countries which are at the root of such calamities.

Food aid, channelled through relief agencies, is of course an immensely important immediate response. However, it has to be recognised that the affected individuals are likely to have multiple nutritional deficiencies, not only of protein and calories, but also of vitamins, minerals and other dietary components. Moreover, the nutritional needs of different groups within the population, such as young children, pregnant women, nursing mothers, manual workers and the elderly, are quite different. Local conditions also vary enormously; parasitic infections are usually rife and resistance to even minor infectious diseases may be very low. Food storage conditions are usually far from ideal, resulting in spoiling of locally-produced and imported foodstuffs alike.

Figure 2.1. Child from a famine area in Ethiopia. Photograph kindly supplied by the Catholic Fund for Overseas Development, London.

Consequently, relief schemes must be devised and coordinated by experts well acquainted with local conditions.

Famine and its consequences are not intractable. Relatively simple measures would make an immense difference if only there was the political and humanitarian will to do it!

The Problem

Protein-rich food supplements were hurriedly despatched to two African famine zones, the arid north-east region of Ethiopia and the humid equatorial region of southern Sudan. Relief workers in Ethiopia reported that about 80% of the children in their region were saved by the food supplements. In contrast, the Sudanese workers found that less than 15% survived and even claimed that the protein appeared to have hastened the children's deaths.

The children in Ethiopia were clearly emaciated (Figure 2.1), whereas the Sudanese children often had a deceptively plump appearance with swollen abdomens and oedematous limbs like the children in Figure 2.2. Table 2.1 shows the serum albumin concentration in groups of children from each region. Some autopsies were performed on the Sudanese casualties and fatty degeneration of the liver was noted. Analysis of the starchy, but limited, local foodstuffs in the Sudan revealed significant contamination by fungal toxins, including aflatoxin B_1. Laboratory tests showed that incubation of aflatoxin B_1 with rat liver microsomes and NADPH produced metabolites which bound to DNA with high affinity.

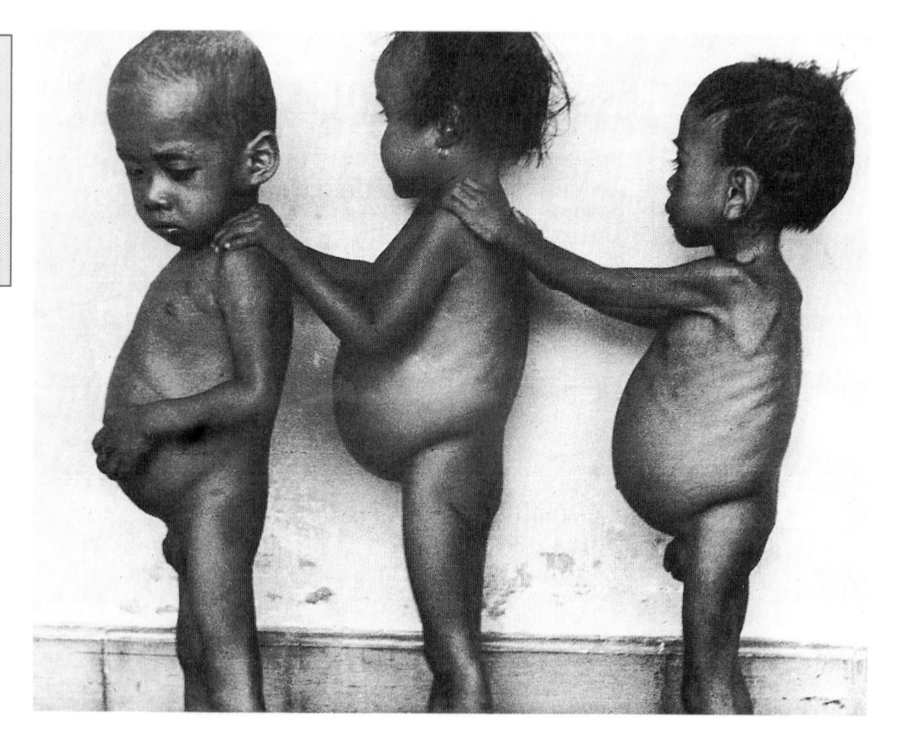

Figure 2.2. Children from a famine area in the tropics. Note the swollen abdomen and oedema of the limbs. Photograph courtesy of the World Health Organization.

Table 2.1. Serum albumin assays.

	Serum albumin (g/l)
Sudanese children	16–25
Ethiopian children	23–29
Well-nourished controls	35–50

Questions

1. What is the basic biochemistry behind the Ethiopian children's emaciated appearance?
2. Why do the Sudanese children have oedema?
3. Consider the ways in which fats are transported around the body. What does the fatty infiltration of the Sudanese children's livers suggest about their liver function?
4. What are microsomes? What do the laboratory tests indicate about the metabolism of aflatoxin B_1 in the liver?
5. What might be the consequences of the aflatoxin metabolites binding to the DNA of liver cells?
6. What could explain the high level of aflatoxin contamination of local foodstuffs in the Sudan?
7. Why was feeding large amounts of protein-rich food paradoxically detrimental to the Sudanese children?
8. How would you recommend the Sudanese relief workers modify their feeding procedures in the future?

Commentary

As a result of the widespread famine, the Ethiopian children are suffering from a severe dietary deficiency of protein and energy foodstuffs known as marasmus (from Greek, meaning 'to waste'). Occurrence of marasmus is not related to climate and used to occur in industrialised areas of Europe.

In the absence of adequate food intake, the body mobilises its fat stores. However, fatty acids cannot be used to maintain blood glucose, which is vital for a variety of organs, particularly the brain, and erythrocytes. Liver glycogen stores are rapidly exhausted, and so muscle protein must be catabolised to provide amino acid carbon sources for gluconeogenesis.

Although the Sudanese children are also severely malnourished, their oedema conceals the underlying muscle wasting and poor growth. Hence, when these children are given high quality food, the first improvement usually seen is a reduction in their oedema, which then reveals the effects of the malnutrition. Among the principal proteins produced by the liver is serum albumin, a major osmotic buffer of extracellular fluid. The marked deficiency of serum albumin in the Sudanese children leads to water retention in the tissues. Another important biosynthetic role for the liver is to provide apolipoproteins for the formation of serum lipoprotein complexes. These are responsible for the transport of lipids, especially triglycerides, around the body. When lipoprotein synthesis is compromised, lipids accumulate in the liver.

Both fatty degeneration of the liver and diminished serum albumin synthesis point to liver failure in the Sudanese children. This is probably the result of ingesting aflatoxins in the local foodstuffs. Aflatoxins, of which aflatoxin B_1 is one of the most potent, are produced by moulds of the genus *Aspergillus*. These xenobiotics are not themselves toxic but their metabolism in the liver produces toxic intermediates. The enzymes responsible for the conversion are associated with the endoplasmic reticulum. The metabolism of aflatoxins can be studied *in vitro* using rat liver microsomes, which are fragments of the endoplasmic reticulum produced during mechanical disruption of liver cells. Hydroxylation of aflatoxin B_1 by the hepatic microsomal P_{450} monooxygenase system in the presence of O_2 and NADPH produces a highly reactive and toxic derivative (actually an aflatoxin epoxide). This may undergo a variety of reactions and thereby disrupt many metabolic processes. In particular, it reacts avidly and covalently with DNA, thus impairing gene expression in the liver. Figure 2.3 shows how aflatoxin B_1 epoxide is thought to bind to guanine residues in DNA.

Ingestion of large amounts of protein results in an excess of amino acids over the immediate requirements for protein synthesis. The surplus amino acids cannot be stored; instead they are catabolised. The unwanted amino groups give rise to ammonia which must be detoxified by conversion to urea. This takes place largely in the liver via the urea cycle. Presumably the Sudanese children's damaged livers have only a limited ability to handle excess amino acids. Thus, feeding large amounts of protein supplements, far from saving the Sudanese children as intended, inadvertently overloaded their weakened livers and contributed to their eventual demise. Smaller amounts of the supplements should be given at more frequent intervals to spread the load on the failing liver.

The Sudanese children are suffering from a form of malnutrition known as kwashiorkor, first described medically by Cecily Williams in the Gold Coast (now Ghana) in 1933. Kwashiorkor is largely restricted to the (sub)tropical regions of west

Figure 2.3. Metabolism of aflatoxin B_1 and the interaction of its epoxide with DNA. The blue structure shows the guanine base in DNA to which the epoxide is linked.

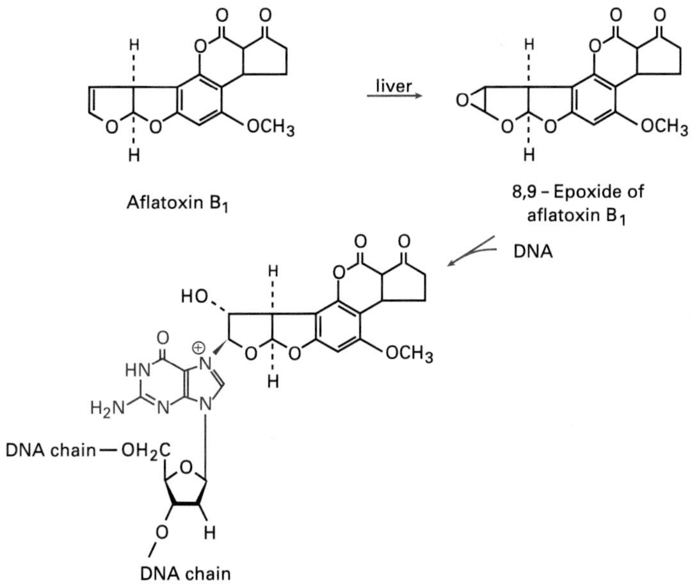

Aflatoxin B_1 → liver → 8,9 - Epoxide of aflatoxin B_1 ← DNA

DNA chain — OH_2C

DNA chain

and central Africa. The word comes from the Ga language of Ghana and loosely translated means 'the sickness of the child who is displaced from the breast'. This is because the condition is frequently associated with the weaning of a child onto an inadequate diet, following the birth of a younger sibling. Kwashiorkor is often said to result from predominantly starchy diets which may be adequate in calories but are grossly deficient in protein. However, protein-energy malnutrition is not the complete story and does not explain kwashiorkor's limited geographical occurrence. While the exact ætiology is still controversial, it is clear that kwashiorkor is complicated by other factors such as the liver damage due to contaminated food, as in the case described here, and various parasitic infections. The hot and humid climate provides ideal conditions for the growth of moulds in foodstuffs, particularly when they are stored under rather primitive conditions in rural areas.

In both marasmus and kwashiorkor the malnutrition leads to a generalised debility and an inability to resist infections, so that death usually results from pneumonia or the consequences of chronic diarrhoea.

Further Questions

1. Malnutrition is not simply a matter of dietary insufficiency of protein and/or calories. Vitamins and minerals will also be limited. Vitamin A deficiency is frequently a complicating factor. What is vitamin A, what are its roles and what are the symptoms of its deficiency? From what you can discover about vitamin A transport in the body, can you explain why vitamin A deficiency is particularly noticeable in kwashiorkor?

2. From the information given in this problem, can you suggest a reason why hepatoma (liver cancer) is one of the most common forms of cancer in Africa but is relatively rare in Europe and USA? (See also Problem 14.)

3. An outbreak of liver disease in chickens, which decimated the poultry industry in Britain in the early 1960s, was caused by aflatoxins in peanuts imported for use in poultry feed. As a result, aflatoxin levels in human and animal foods are now routinely monitored. Can you use the information that you have gained in this problem to devise a biochemical test system suitable for this purpose?
4. Which hormones are involved in controlling energy metabolism in starvation? (See also Problems 1, 12 and 18.)
5. Why are fatty acids not gluconeogenic?
6. In a hitherto well-nourished individual who starves, why does muscle catabolism take place at a much higher rate in the first few days than later?
7. Chronic diarrhoea can be fatal. Why is this and how would you treat it? (See also Problem 13.)

Connections

- Summarise the components required in a healthy diet. Do you understand the terms *nitrogen balance, vitamin, essential amino acid, essential fatty acid*? What are some of the most prevalent dietary deficiency disorders? (See also Problem 6.)
- Review the central role of the liver in metabolism and the integration of the major metabolic pathways. (See also Problems 1, 5 and 19.)
- Review what you know about 'detoxification mechanisms' from the point of view of xenobiotics and endogenously-produced 'toxins' such as bilirubin. (See also Problems 14 and 17.)
- Recall what roles the liver P_{450} monooxygenase system has in biosynthesis. (See also Problem 7.)
- Check that you know what proteins occur in plasma, what they do and how you would separate and identify them. (See also Problems 3 and 20.)
- Ensure you understand how osmotic homeostasis is maintained. (See also Problem 13.)
- Summarise what you know about the structures and composition of serum lipoproteins and their roles in lipid transport. (See also Problem 20.)

References

Hendrickse R G (1991) Kwashiorkor: the hypothesis that incriminates aflatoxins. *Pediatrics* **88**, 376–9.

Marsden P D (1990) Kwashiorkor. *British Medical Journal*, **301**, 1036–7. Good description of a Brazilian case.

Read C (1990) Behind the face of malnutrition. *New Scientist*, 17 February 1990, pp 38–42.

3 CHEST PAINS

Introduction

Cardiovascular disease is a major cause of death in Western society (Figure 3.1) with coronary heart disease being the single most important cause of death of middle-aged men. When a fibrin clot occludes a branch of the coronary arterial 'tree', a region of the heart muscle (myocardium) becomes starved of blood (ischaemic). The ischaemic myocardium then undergoes necrosis and a myocardial infarction is said to have taken place. Surviving a heart attack depends on prompt medical intervention. However, the symptoms of myocardial infarction are similar to those of other less threatening conditions, so that accurate diagnosis is essential to ensure appropriate follow-up treatment.

Lactate dehydrogenase (LDH) and creatine kinase (CK) are two enzymes particularly useful in diagnosing whether a myocardial infarction has occurred. LDH is a widely distributed tetrameric enzyme. There are two types of subunit (M and H) which in combination give five isoenzymes, separable by electrophoresis. Each isoenzyme has a characteristic tissue distribution (Table 3.1).

Creatine kinase is a dimeric protein which also has two types of subunit (M and B). Its isoenzymes also show a characteristic tissue distribution (Table 3.2).

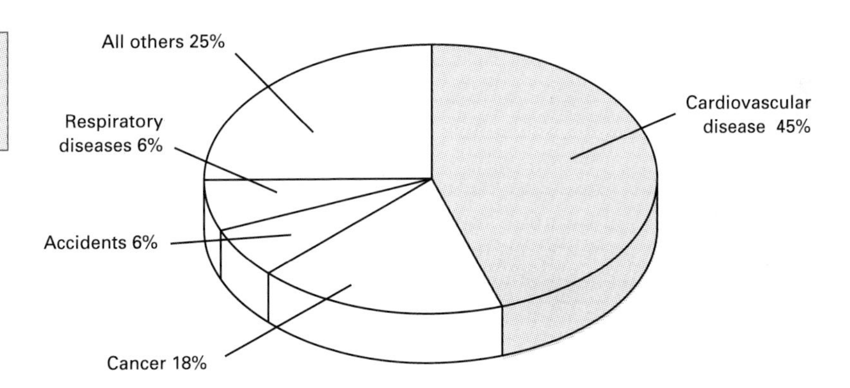

Figure 3.1. Major causes of death in Western society.

Table 3.1 Tissue distribution of LDH isoenzymes.

Isoenzyme	Subunit composition	Tissues in which the isoenzyme occurs as a major form
LDH1	HHHH	Myocardium, erythrocytes
LDH2	HHHM	Myocardium, erythrocytes
LDH3	HHMM	Lymphoid cells, brain, kidney
LDH4	HMMM	Liver, skeletal muscle
LDH5	MMMM	Liver, skeletal muscle

Table 3.2. Tissue distribution of creatine kinase isoenzymes.

	Isoenzyme		
Tissue	MM	MB	BB
Skeletal muscle	95%	2–3%	1–2%
Myocardium	80–85%	15–20%	1%
Brain	10%	1–2%	90%

Table 3.3. Serum enzyme activities (U/l) in the patients.

Enzyme activity	Ralph K.		George P.		Reference values (range)
	10 hours	24 hours	10 hours	24 hours	
Total CK	748	850	735	790	20–150
CK-MB	82	75	25	23	0–3
LDH	129	182	106	167	45–90

The Problem

You are in charge of a coronary care unit. The previous day, there were two admissions via the Accident and Emergency Service of patients with suspected myocardial infarction. Both men were in their late forties and had suffered intense chest pains. Ralph K., a businessman, had been taken ill at a local restaurant, while George P., a teacher, had just returned home after a strenuous jogging session. Two blood samples had been taken from each patient, at approximately 10 hours and 24 hours after the onset of the chest pains, for assay of serum CK and LDH (Table 3.3 and Figure 3.2). Now you are faced with another acute admission and there is no available bed in the coronary unit.

Questions

1. In general terms, what is the rationale behind the measurement of serum enzymes in these and other patients?

2. In taking and processing the blood samples, what precautions must be observed to ensure that the results will be valid and informative?

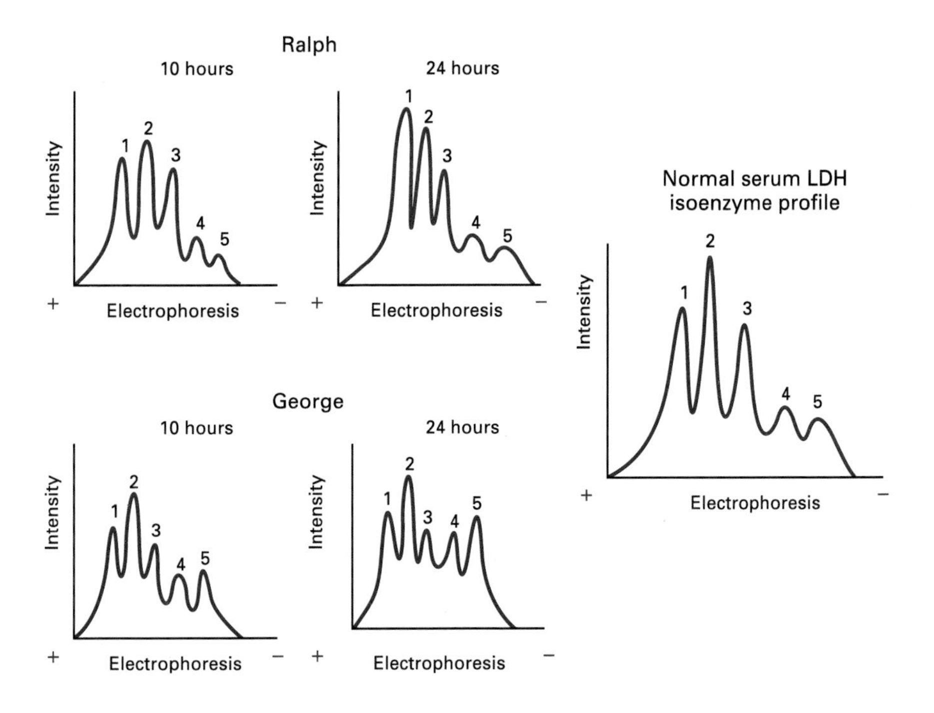

Figure 3.2. Electrophoresis of LDH isoenzymes. The serum was subjected to electrophoresis on cellulose acetate membranes and then the membranes were treated to reveal 'bands' of LDH activity. The activity in each 'band' was determined by scanning with a photoelectric densitometer. (The numbers refer to the LDH isoenzymes in Table 3.1. Intensity = intensity of staining.)

3. CK-MB was assayed by immuno-inhibition. After measuring the total CK, an excess of an antibody specific for the M subunit of CK was added to each serum sample. The CK activity remaining was compared with the total CK activity to arrive at a value for CK-MB. What is the rationale of this determination? What assumption has to be made? From the information presented in this problem, do you think this assumption is valid?

4. Bearing in mind that an initial myocardial infarction may be shortly followed by another, is it possible from the serum enzyme data to decide whether either Ralph or George could be moved to a medical ward to make room in your coronary care unit for the latest admission? What are the deciding factors?

5. Would screening the population at large for serum enzymes such as CK and LDH be of predictive value in identifying people at risk of heart attack?

6. One of the treatments that may be given in this situation is the intravenous administration of tissue-type plasminogen activator (tPA). What is the basis of this treatment?

Commentary

Cytosolic proteins, including CK and LDH, are normally efficiently retained inside cells by the selectively permeable plasma membrane. Tissue damage may compromise membrane integrity sufficiently to allow release of these cytosolic proteins. The insult does not necessarily have to be large enough to result in cell death. Release of tissue proteins into the extracellular fluid may occur because of exposure to toxic chemicals, for example in cirrhosis of the liver due to alcohol abuse. Any significant impairment in the cell's energy status may also compromise membrane integrity since this is intimately bound up with the actions of various energy-driven pumps in the cell membrane. Ischaemia, the temporary interruption in tissue blood supply, results in partial tissue anoxia and interferes with energy production.

While some measure of the overall scale of the tissue damage can be gained from the rise in *total* enzyme activities, the particular tissues involved cannot usually be pinpointed since many of the most conveniently assayed enzymes are present in a variety of different tissues. However, *isoenzymes* often allow a differential diagnosis to be made since their distribution is tissue-specific, as illustrated in this problem by CK and LDH isoenzymes.

Several methods can be used to separate and quantify isoenzymes. These include electrophoresis (used here for LDH), various types of chromatography, differential heat stability, sensitivity to inhibitors, or immunological methods. The method of choice depends on the particular enzyme but also on the speed, sensitivity and specificity of the method. In the case of CK-MB, immuno-inhibition currently offers the best method. The antibody binds to and inhibits the activity of M subunits, leaving the activity of B subunits unaffected. Remaining CK activity is assumed to be due only to CK-MB since significant amounts of CK-BB are rarely found in serum; presumably the blood–brain barrier prevents it entering the general circulation from the brain, its only major source.

It should be remembered that erythrocytes often contain large amounts of enzymes used in diagnosis, for instance LDH, so it is vital that haemolysis is avoided both when the blood sample is taken and during storage of the sample before assay. In this case you are faced with deciding whether either of the patients really has suffered a myocardial infarction, necessitating his continued stay in intensive care. Both patients have obviously incurred severe muscle damage resulting in the release of both CK and LDH. But has this involved skeletal muscle or cardiac muscle or both? The CK-MB isoenzyme pattern is particularly helpful in the differential diagnosis. Even though this isoenzyme occurs in both skeletal and cardiac muscle, the proportions are quite different. Regardless of the extent of muscle damage, if CK-MB constitutes significantly more than 3% of the total CK, myocardial involvement can be suspected. This is clearly the case for Ralph (11% at 10 hours and 9% at 24 hours). George's CK-MB level remains at about 3% suggesting skeletal muscle damage (too much jogging!) rather than a myocardial infarction as the source of his serum CK. This diagnosis is supported by the LDH isoenzyme patterns where the skeletal muscle forms (LDH4 and LDH5) are disproportionately increased in George's serum. In Ralph, it is the cardiac isoenzymes (LDH1 and LDH2) that are elevated; in fact the normal ratio LDH1:LDH2 is reversed in the 24 hour sample, a characteristic sign of myocardial infarction. Generally the elevations in total CK, and CK-MB in particular, exceed and precede changes in LDH and its isoenzymes. CK therefore provides an earlier confirmation of myocardial infarction and LDH would be used for confirmation. So it

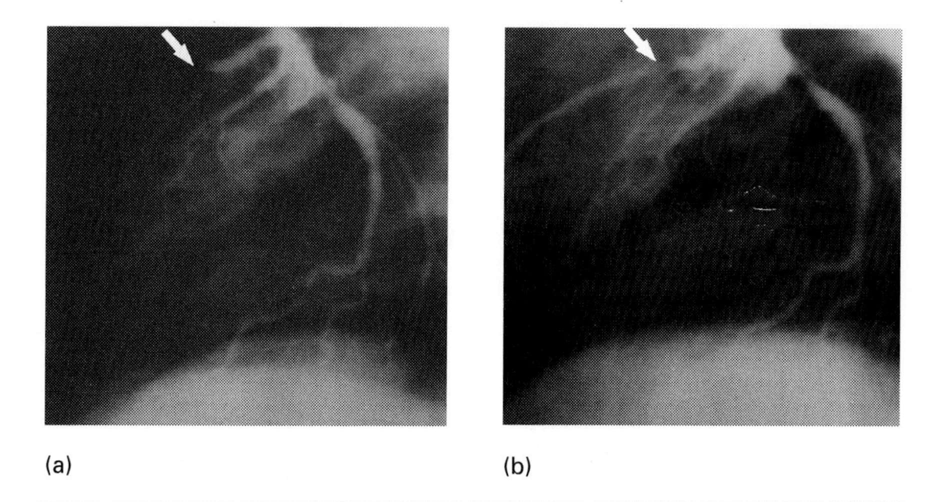

Figure 3.3. Coronary angiograms from a patient before (panel a) and after (panel b) thrombolysis with intravenous tPA showing restoration of patency in the left anterior descending coronary artery (arrow). Photograph kindly supplied by Dr Burton Sobel (Washington University School of Medicine, St Louis, USA) and reproduced from Van de Werf *et al.* (1984) *New England Journal of Medicine* **310**, 609–13, with permission.

is probably safe to discharge George from intensive care in order to admit the new case.

It is important to remember that enzyme changes such as these, occurring as they do as a *consequence* of tissue damage, can only ever be witnesses to earlier events. For this reason, it would not be useful to screen the general population for elevations in CK or LDH since they would be of little *predictive* value. Nevertheless, as this case illustrates, enzyme analysis can be of benefit in establishing whether or not a myocardial infarction has occurred and ensuring appropriate patient management.

Blood clots are normally dissolved by the enzyme plasmin which is formed from its inactive precursor plasminogen by the action of tissue-type plasminogen activator (tPA). Now that the gene for tPA has been cloned, recombinant tPA is available for treating patients with a view to dissolving blood clots *in situ* (see Figure 3.3). Streptokinase, an enzyme prepared from haemolytic streptococci, will also activate plasminogen and is a cheaper alternative to tPA.

Further Questions

1. If the serum LDH isoenzyme pattern shown in Figure 3.4 had been obtained, which tissue(s) might have been involved?
2. What would the serum LDH isoenzyme pattern look like in a case of liver damage?
3. The following CK activity measurements were obtained using the immuno-inhibition method: before antibody = 400 U/l; after antibody = 20 U/l. How much

Figure 3.4. LDH isoenzymes in another patient.

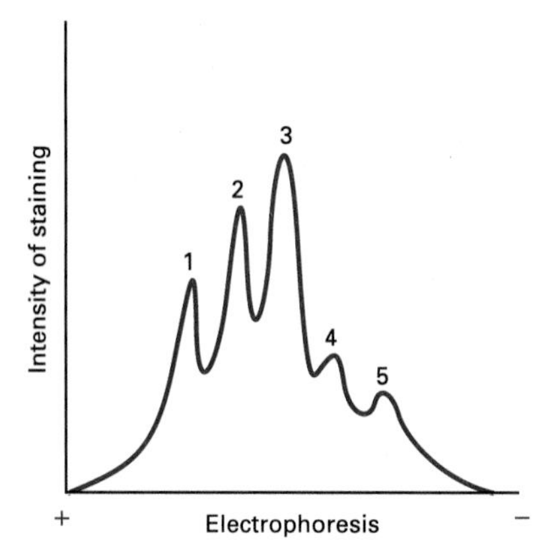

CK-MB is present, assuming there is no CK-BB? (See also Problem 12 for another type of assay involving antibodies.)

4. What reasons can you suggest for different tissues expressing different isoenzyme patterns? To help you answer this question, it may be helpful if you review the different kinetic properties and roles of glucokinase (K_m for glucose ~10 mM) and hexokinase (K_m for glucose 0.01–0.1 mM), which may, in a sense, be regarded as isoenzymes.

5. How would you go about preparing an antibody specific for the M subunit of CK and then show that it inhibited the appropriate enzyme activity? For the type of assay featured in this problem, would a polyclonal or a monoclonal antibody preparation be appropriate?

6. In this problem enzyme *activities* were measured. How would you measure the *amount* of LDH present in a serum sample?

Connections

- Review your knowledge of membrane structure and the various transport processes that enable membranes to be selectively permeable. Distinguish between active and passive transport. (See also Problem 13.)
- Use this problem to revise the types of metabolic fuels available to muscle for generating energy. Do you understand why the energy yields are different? Review energy metabolism in starvation and during uncontrolled diabetes mellitus. (See also Problems 1 and 2.)
- Review your knowledge of the process of blood coagulation. What other biological cascades do you know? Can you think of anticoagulant therapies other than treatment with tPA or streptokinase?
- Use this problem to refresh your memory as to how protein electrophoresis may be carried out. Appreciate that it can be done with native or denatured proteins; which would have been appropriate in this problem? Describe methods that could

be used to locate the protein 'bands' after the electrophoresis. (See also Problems 8, 11, 18, 20 and 22 for other examples of protein electrophoresis.)

- Take this opportunity to remember the different uses to which enzymes can be put in clinical diagnosis: indicating tissue damage, pinpointing genetic defects, assaying key metabolites in cell extracts or body fluids – what others can you recall?

References

Marshall W J (1992) *Clinical Chemistry* (2nd edn), Clinical Enzymology, Chapter 17. Gower Medical Publishing, London.

Mercer D W and Varat M A (1987) Creatine kinase isoenzymes and the diagnosis of myocardial infarction, Chapter 1. In *Clinical Studies in Medical Biochemistry* (Glew R M & Peters S P, eds). Oxford University Press, Oxford and New York.

4 DEATH ON THE FARM

Introduction

The success and profitability of modern agriculture depend on, among other things, tight control of pests so that maximal yields of crops are obtained. To some extent this has been achieved by growing varieties which are naturally resistant to pests. With the advent of the 'biotechnological revolution', we are now moving into the era of 'designer plants' and there will be increased reliance on crops genetically engineered for resistance to specific pests. However, in the meantime, agriculture relies heavily on spraying crops with chemical pesticides.

The ideal pesticide is of course one that is selectively toxic to the pest but harmless to humans and other species. In fact a great deal of effort is put into discovering such chemicals, but rarely is this goal of specificity attained. The difficulty is that eukaryotic species have so much in common that there are very few differences in their metabolism which can be the focal points for differential attack. This problem is not just confined to pest control in agriculture; in treating patients with cancer or with diseases due to eukaryotic parasites, the clinician faces the dilemma of knowing that the drugs used in chemotherapy will often also harm the patient.

Because of this, agents are used in agriculture that are substantially toxic to humans and therefore hazardous to agricultural workers, particularly during crop spraying, unless stringent precautions are taken (Figure 4.1).

Figure 4.1. Agricultural workers spraying crops.

The Problem

Two agricultural workers were employed to spray cereal crops with a preparation containing *o*-dinitrocresol (DNOC) for pest control. On the final day of the spraying programme both men began to feel ill, and one of them died whilst returning to the farm for help. The doctor who examined the corpse shortly afterwards noted that rigor mortis was unusually far advanced. The other man was admitted to hospital. His temperature was 39.3°C and he was sweating profusely with a respiratory rate of 60/min. He was extremely anxious and his pupils were dilated. Despite strenuous attempts to reduce his body temperature, he also died in a coma shortly after admission. Autopsy findings included pulmonary congestion and slight oedema. There was a striking absence of subcutaneous and omental fat. The shaft of the femur contained red marrow throughout, with microscopic findings consistent with anoxaemia. In view of the clinical findings, a sample of the pesticide was tested on a preparation of rat liver mitochondria. Representative recordings obtained with an oxygen electrode are shown in Figure 4.2.

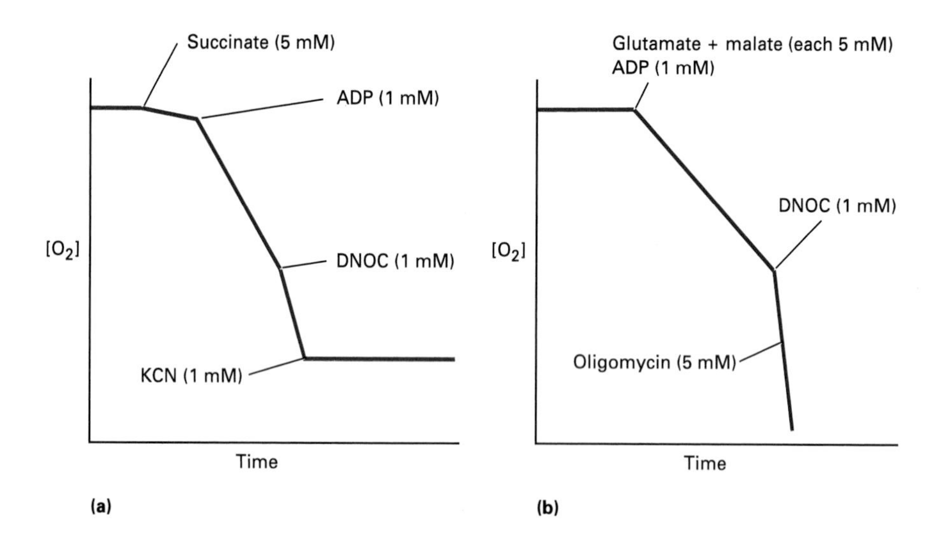

Figure 4.2. Effects of DNOC on rat liver mitochondria. A suspension of freshly-prepared rat liver mitochondria in a suitable buffer was incubated in an oxygen electrode chamber. Various compounds were added to the chamber as indicated (chamber concentrations shown) and changes in the oxygen consumption were recorded.

Questions

1. Draw a simple diagram of the mitochondrial electron transport system to indicate the order of the principal redox carriers. Where are the sites of oxidative phosphorylation and the entry points for the substrates used in the oxygen electrode experiments?
2. Why is mitochondrial respiration dependent on ADP (Figure 4.2a)? What is the reason for adding both glutamate and malate (Figure 4.2b)? Why is the rate of respiration in the presence of succinate and ADP (Figure 4.2a) greater than with glutamate, malate and ADP (Figure 4.2b)? Explain the results of adding KCN (Figure 4.2a).
3. Under normal circumstances, what would happen after addition of oligomycin (Figure 4.2b)?
4. What do the data allow you to conclude regarding the metabolic effect of DNOC?
5. Why were the respiratory rate and body temperature elevated?
6. Can you provide explanations for the absence of body fat, the effect on the femur and the rapid onset of rigor mortis?

Commentary

Foods are oxidised to release their stored chemical energy. Oxidation involves transfer of electrons from food substrates to O_2 which is reduced to H_2O. Electrons are not passed directly to O_2 but via a chain of reduction–oxidation (redox) carriers (the electron transport chain) located in the mitochondrial inner membrane (Figure 4.3). The chain consists of four multi-component complexes (I, II, III and IV) linked by mobile redox carriers (CoQ and cytochrome *c*). Each complex uses the energy released in the electron transfer to transport H^+ across the membrane, so the $[H^+]$ is higher on the outer side of the membrane. This electrochemical gradient of H^+ represents a temporary store of energy that is used by ATP synthase to phosphorylate ADP to ATP, the H^+ flowing back across the membrane in the process.

Metabolic coupling of the electron transport chain and oxidative phosphorylation ensures that substrates are only metabolised when there is a demand for ATP. The fact that O_2 consumption only occurs at a significant rate when both a substrate and ADP are present (Figure 4.2) indicates that the mitochondria are tightly coupled. Malate enters the mitochondrion and is oxidised to oxaloacetate by malate dehydrogenase with concomitant reduction of mitochondrial NAD^+ to NADH which feeds into the electron transport chain. For continued oxidation of malate, oxaloacetate must exit the mitochondrion. However, oxaloacetate itself is only poorly transported so it is transaminated to aspartate, which is readily transported. The transaminase involved, glutamic-oxaloacetic transaminase (also called aspartate transaminase), utilises glutamate and α-ketoglutarate as the other amino acid/keto acid pair:

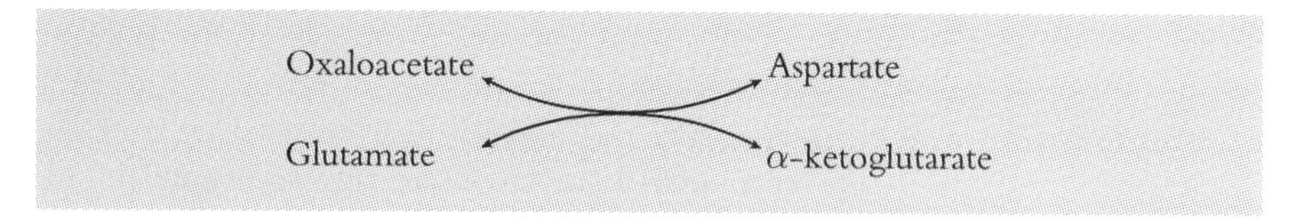

In effect, the oxygen electrode experiment uses part of the mitochondrial aspartate–malate shuttle, a device for allowing NADH produced in the cytosol by glycolysis to be transported indirectly into the mitochondrion for oxidation back to NAD^+.

By reacting covalently with the Fe^{3+} in cytochrome oxidase, the respiratory poison KCN inhibits the terminal step in the electron transport chain, so respiration ceases.

Figure 4.3. Diagrammatic representation of the electron transport chain.

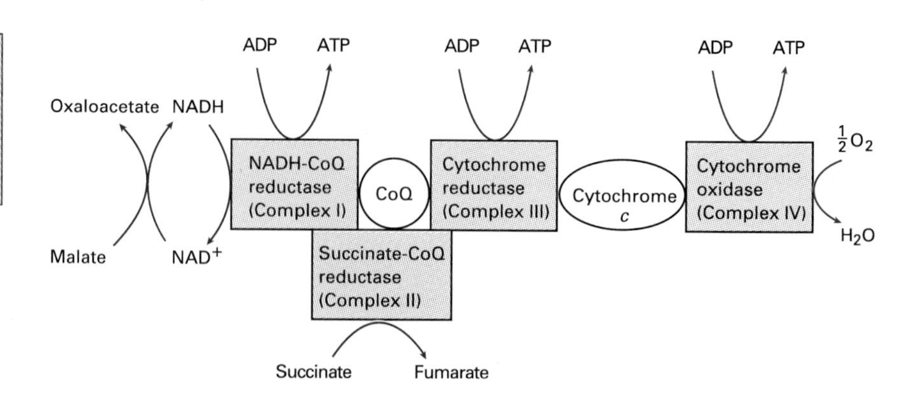

NADH-linked substrates feed electrons into the chain via NADH dehydrogenase (part of Complex I) and so can generate 3 ATP for each oxygen atom reduced (P:O = 3.0). Oxidation of succinate feeds electrons into the chain via $FADH_2$ and succinate dehydrogenase (part of Complex II) thus utilising only two of the phosphorylation sites (P:O = 2.0). So to generate a given quantity of ATP, more succinate must be oxidised than malate and glutamate.

Oligomycin interferes with the ability of the ATP synthase to utilise the H^+ electrochemical gradient. So in coupled mitochondria, oligomycin would be expected to inhibit respiration. However, in this instance oligomycin is without effect suggesting that the prior addition of DNOC has uncoupled the mitochondria. Consistent with this is the large increase in mitochondrial O_2 consumption after addition of DNOC.

Aromatic weak acids such as DNOC and dinitrophenol are thought to pass readily across the mitochondrial inner membrane in their undissociated form thus dissipating the electrochemical gradient (Figure 4.4).

Figure 4.4. Dissipation of the electrochemical gradient by DNOC.

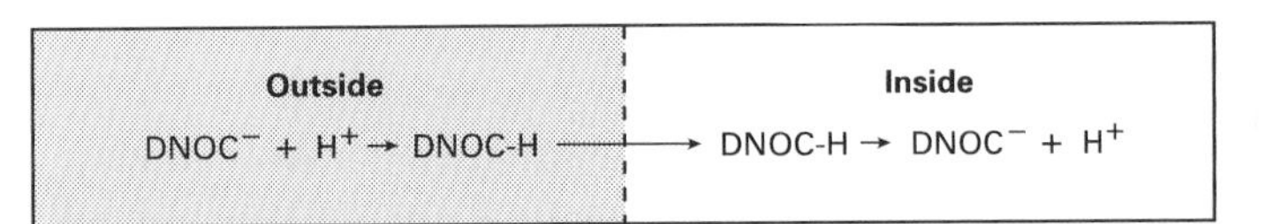

DNOC acts as a pesticide by uncoupling mitochondrial respiration in insects so that their electron transport chain runs uncontrollably and unproductively. The same has happened in the two agricultural workers. Large amounts of metabolic fuels are consumed with the released energy being wasted as heat. The principal fuels used for this uncontrolled respiration are fatty acids from the triglycerides stored in adipose tissue, thus depleting the body's fat stores. The accompanying excessive oxygen consumption leads to tissue hypoxia, which the body attempts to alleviate by increased pulmonary respiration and by erythropoiesis in the bone marrow.

The rigor mortis can be explained by considering muscle biochemistry. The power stroke moves muscle actin filaments relative to the 'heads' of the myosin, so shortening the muscle fibre. This involves ATP hydrolysis to ADP by myosin ATPase. To relax the fibre for the next power stroke, the ADP must be displaced by incoming ATP. Since DNOC poisoning greatly decreases the concentration of ATP, the contractile system is left in rigor.

Further Questions

1. What other mitochondrial shuttles do you know and what are their roles?
2. What are the 'benefits' of segregating the electron transport chain in the mitochondrial inner membrane?
3. Under some circumstances the electron transport chain may be uncoupled from oxidative phosphorylation. At first sight this appears to 'waste' energy but it may have

a physiological role. In which tissue(s) does this type of uncoupling occur and in what circumstances?

4. So-called 'futile cycles', such as that occurring by the simultaneous operation of phosphofructokinase and fructose bisphosphatase, also appear to 'waste' energy by fruitlessly consuming ATP. What are thought to be the advantages of using ATP in this manner?

5. What would have occurred if rotenone or antimycin had been added along with glutamate and malate, or succinate, in Figure 4.2?

6. Many years ago uncouplers such as dinitrophenol were suggested as drugs to aid slimming. What was the rationale and would there be any dangers?

7. In this problem the victim's tissues became anoxic and the response was increased erythropoiesis. How does the body sense anoxia and how is erythropoiesis stimulated? (See also Problems 6 and 11.)

Connections

- Use this problem to review your understanding of mitochondrial oxidative metabolism to produce energy in the cell. This will include a knowledge of the nature of the electron carriers in the electron transport chain, reduction–oxidation potentials and the mechanism of oxidative phosphorylation. Ensure you understand how an oxygen electrode is used to study respiration, how the order of the electron transport carriers was determined and what is meant by the term P:O ratio. (See also Problem 21.)

- Review the mechanism of muscular contraction – the role of the proteins, actin and myosin, the way in which ATP is used to supply energy for contraction and the importance of calcium. Remember that there are many other types of cellular movement – make a list of these and ensure you understand the biochemical principles involved.

- The cytochromes in the electron transport chain are haem proteins, but so are haemoglobin and myoglobin. Remind yourself of the essential differences in the roles of the haem groups in each of these proteins. Respiratory poisons such as cyanide and carbon monoxide inhibit these proteins. Check that you understand the way in which carbon monoxide from a car exhaust causes death. Studies on haemoglobin have told us a lot about how the structure of a protein determines its functions. Make a list of the more important haemoglobin mutations and what they tell us about this structure–function relationship. (See also Problem 11.)

- Although oxidative phosphorylation is by far the most important source of ATP in the majority of cells, ATP can also be produced under anaerobic conditions, as happens during a sprint. Outline how ATP is produced in muscles working under these conditions and what the problems are in sustaining this type of activity. (See also Problem 1.)

- Carbohydrates and fats are the metabolic fuels of the body and they are eventually broken down into compounds which can undergo complete oxidation to CO_2 in the mitochondria. However, they are not exactly equivalent fuels and each has its own advantages and disadvantages. Make a list of these.

- Remind yourself of how mitochondria are prepared from tissue homogenates by subcellular fractionation. How would you ascertain the purity and activity of a preparation of mitochondria?
- NADH is a coenzyme whereas the haem group in a cytochrome is a prosthetic group. Remind yourself of the mechanistic difference between these two terms. NADH and $FADH_2$ are derived from vitamins. Use this to revise your knowledge of the roles of all the vitamins required by humans.
- In this problem NADH produced during oxidation of fuels was involved in the generation of ATP. NADPH is utilised for quite different purposes. Remind yourself of what these are and where NADPH is produced in the cell.
- Many pesticides are carcinogenic. Check that you understand how carcinogens act as mutagens, and the relationship between carcinogens and pre-carcinogens. (See also Problem 14.)

Reference

Council of Scientific Affairs report. Cancer risk of pesticides in agricultural workers. *Journal of the American Medical Association* **260**, 959–66.

5 FAILURE TO THRIVE

Introduction

Mutations which affect enzymes of the body's metabolic pathways are often referred to as 'inborn errors of metabolism', a term coined by Sir Archibald Garrod at the turn of the 20th Century. Garrod's work on the rare disorder of tyrosine degradation, alkaptonuria, laid the foundation of modern biochemical genetics.

An inborn error of metabolism arises because a mutation affects the activity of an enzyme in a metabolic pathway. Mutations may be of several kinds such that the enzyme protein may be entirely lacking or, more usually, produced in altered form with diminished catalytic activity.

The precise clinical picture which emerges in an inborn error of metabolism is usually very complicated since it depends on many factors. In the heterozygous condition, there is often sufficient enzyme produced by the normal gene to mask the deleterious effects of the mutation. The normal gene is said to have a functional reserve and this is a common finding with inborn errors. However, in some cases this functional reserve is lacking and the heterozygote will also display the inborn error, for instance in acute intermittent porphyria. Yet again, where the gene involved is carried on the X chromosome rather than on an autosome, the error will be seen predominantly in males; the female usually only shows the disorder in the homozygous condition. An example of this would be glucose 6-phosphate dehydrogenase deficiency.

Another factor to be taken into account when trying to explain the phenotype associated with an inborn error of metabolism is the extent to which the affected pathway is vital for life. Presumably the reason why inborn errors of the tricarboxylic acid cycle are rarely seen reflects the central role of this cycle in metabolism. Affected individuals would be expected to die *in utero*. Blockage of a pathway need not of course be total; many mutations are 'leaky' in that an altered enzyme is produced which does have some residual activity. Finally, it must be remembered that during foetal life many metabolic activities are handled by the mother and only become the responsibility of the child after birth.

Inborn errors of metabolism are just one class of monogenic disorder. Although a few monogenic disorders are common enough to warrant genetic screening programmes, such as that for phenylketonuria, most are rare (see Appendix 4). However, many of the major diseases affecting the developed world have a significant but complex genetic component. These polygenic disorders include diabetes mellitus, cardiovascular disease and hypertension, schizophrenia, epilepsy, depression, neural tube defects and some forms of cancer. Collectively, genetic disorders represent a considerable 'burden' in the human population (very approximately estimated at perhaps 5% of live births).

The Problem

A male child, John S., had been delivered normally after an uncomplicated pregnancy. For about a week his condition appeared to be normal, but he then became progressively more lethargic, failed to thrive and had recurrent bouts of irritability and vomiting. He was admitted to hospital in a distressed condition and a blood test revealed the data shown in Table 5.1. Shortly after admission to hospital, John's condition worsened and he became comatose. Treatment was initiated and his life was saved.

Table 5.1 Patient's blood analysis.

	John S.	Normal range
pH	7.55	7.35–7.45
urea	0.7 mM	2.5–7.0 mM
NH_3	420 μM	12–60 μM
glutamine	1.5 mM	0.45–0.75 mM
citrulline	1.2 mM	10–30 μM
argininosuccinic acid	none detected	none detectable

Questions

1. Consider the blood analysis (Table 5.1). What is the biochemical lesion in the patient and how does it account for the blood changes?
2. What might have been responsible for the lethargy, irritability and vomiting which prompted John's initial admission to hospital?
3. What immediate treatment would have been given to save his life?
4. Why did the condition only become apparent several days after birth and not earlier or *in utero*?
5. What long-term therapy would be appropriate and how would its success be monitored?
6. The long-term treatment actually given included oral arginine and sodium phenylacetate. What is the rationale of this treatment (the metabolism of the latter compound results in the excretion of phenylacetylglutamine)?

Commentary

The data in Table 5.1 show that this patient suffers from hyperammonaemia. Ammonia, most of which is present as NH_4^+, is extremely toxic especially to the central nervous system, hence the irritability, lethargy, vomiting and eventually the coma. Why ammonia is neurotoxic is not clear. However, the brain is particularly energy demanding. In the presence of excess NH_4^+, the glutamine synthetase reaction favours the formation of glutamine:

$$\text{Glutamate} + NH_4^+ + \text{ATP} \rightarrow \text{Glutamine} + \text{ADP} + P_i + H^+$$

In addition to using up ATP, this reaction will utilise glutamate which has to be replenished by transamination of α-ketoglutarate, in turn depleting this crucial tricarboxylic acid cycle intermediate. Glutamate itself is the major excitatory neurotransmitter in the brain. Some or all of these mechanisms may interfere with brain metabolism.

Excess amino acids derived from dietary protein cannot be stored as such. Their carbon skeletons can be stored in the form of carbohydrate and/or lipid, but their amino groups are converted into NH_4^+. Humans, like most mammals, are ureotelic excreting NH_4^+ in the form of urea, a relatively non-toxic water-soluble compound. Urea synthesis occurs chiefly in the liver via the urea cycle (Figure 5.1). Discovered in the early 1930s by Hans Krebs and Kurt Henseleit (a medical student), the urea cycle dramatically changed biochemists' thinking away from simple linear metabolic pathways. Since then, several other metabolic cycles have been found, most notably the tricarboxylic acid cycle, the discovery of which earned Hans Krebs his Nobel Prize.

The defect in the patient is pinpointed to the argininosuccinate synthetase step of the urea cycle (Figure 5.1) by the elevated citrulline but absence of argininosuccinate (Table 5.1). The deficiency prevents adequate urea synthesis and disposal of NH_4^+, though the deficiency is clearly not complete. Deficiencies in other enzymes of the

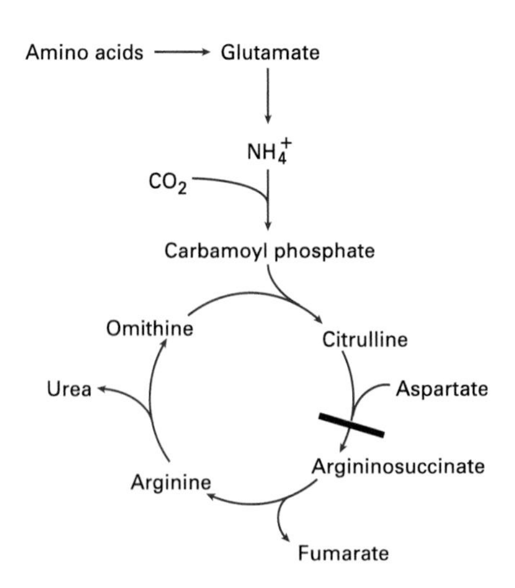

Figure 5.1. The urea cycle showing the metabolic block in the patient.

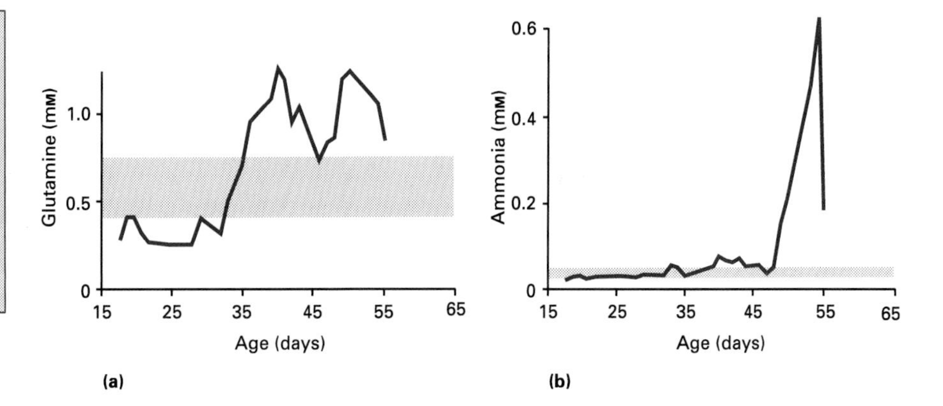

Figure 5.2. The value of monitoring the plasma glutamine levels (a) to give a warning of impending hyperammonaemia (b). The shaded areas indicate the normal plasma ranges for glutamine and NH_4^+.

urea cycle are also known but collectively they are rather rare (1 in 25 000 live births). They only become manifest postnatally since foetal NH_4^+ is handled by the maternal liver. Urea cycle enzymes are induced hormonally after birth and any genetic deficiency affecting the urea cycle takes at least 1–2 days to become apparent.

The immediate threat to this patient was the hyperammonaemic coma. John needed haemodialysis to reduce his plasma NH_4^+ quickly to normal. Thereafter, it must be stabilised within the normal range, otherwise irreversible brain damage will ensue. John's protein intake has to be restricted and given in frequent small amounts to prevent his inadequate urea cycle being overloaded. However, enough protein must be available to ensure adequate growth. At the same time, alternative routes for nitrogen excretion must be provided. Phenylacetate and arginine are designed to do this. Glutamine combines with phenylacetate and is excreted, along with its nitrogen, as phenylacetylglutamine. The arginine therapy exploits part of the urea cycle; arginine is metabolised as far as citrulline, on the way picking up NH_4^+ in the form of carbamoyl phosphate, and the accumulated citrulline is excreted.

The success of the treatment depends on early warning of impending hyperammonaemia. This is provided by monitoring blood glutamine (Figure 5.2) since the glutamine synthetase system buffers NH_4^+. With this type of therapy, the prospects are now much better than hitherto when only 10% of patients could be expected to survive one year beyond diagnosis. However, even though patients with urea cycle disorders now survive they are not free of problems and many are retarded, presumably due to the effects of the first few days of hyperammonaemia before diagnosis and commencement of therapy.

Further Questions

1. 'Jack Spratt could eat no fat, his wife could eat no lean.' So runs the nursery rhyme. Mrs Spratt may have been heterozygous for another urea cycle defect, a deficiency of ornithine transcarbamoylase, which is an X-linked disorder. In what ways would Mrs Spratt's symptoms resemble those of John S. and how would they differ? Why might

she excrete large amounts of orotic acid? Pregnancy would be a particularly difficult time for Mrs Spratt, why?

2. The kidney normally produces and excretes NH_4^+. In conditions of metabolic acidosis, e.g. the ketoacidosis of starvation, this NH_4^+ excretion increases. How is the NH_4^+ produced and what is its role?
3. How much protein nitrogen can John excrete (in g/kg/day) while he is being treated with arginine and phenylacetate (each 0.5 g/kg/day)?
4. John could have been treated with sodium benzoate in place of phenylacetate. Benzoate combines with glycine to be excreted as hippurate. Why is benzoate less effective than phenylacetate?

Connections

- Use this problem to review your knowledge of the major metabolic pathways, such as glycolysis, gluconeogenesis, the tricarboxylic acid cycle and β-oxidation of fatty acids. Ensure you understand how these are integrated together and controlled.
- Check that you understand how amino acids are transaminated. What vitamin is involved? Why are some amino acids 'essential'? (See also Problems 10 and 19.)
- Check that you can name examples of glucogenic and ketogenic amino acids and explain the difference.
- Remind yourself of how plasma pH is regulated and the roles of the various buffer systems, the lungs and the kidney. (See also Problem 10.)
- Other nitrogen excretion products besides urea are uric acid and creatinine. Check that you know what each of these is and how it is formed. Review what you know about gout.
- Urinary excretion of nitrogen compounds reflects the dietary status of the subject. Consider three types of dietary regime – very high protein with adequate calories, very low protein but with adequate calories, and minimal food intake. Make a list of the principal changes which you would expect to see in the urinary analysis and then check that you can explain them.
- List other inborn errors of metabolism which you have come across. Look up what the defect is and see how far the biochemistry can explain the symptoms. (See also Problems 1, 7, 10 and 19.)

References

Arn P H, Hauser E R, Thomas G H, Herman G, Hess D and Brusilow S W (1990) Hyperammonemia in women with a mutation at the ornithine transcarbamoyl-transferase locus. *New England Journal of Medicine* **322**, 1652–5.

Grompe K, Jones S N and Caskey C T (1990) Molecular detection and correction of ornithine transcarbamylase deficiency. *Trends in Genetics* **6**, 335–9.

Watford M (1991) The urea cycle: a two-compartment system. *Essays in Biochemistry* **26**, 49–58.

6 THIN BLOOD

Introduction

To a clinician the results of a blood test are often very important to the diagnosis. A blood test is obviously crucial in cases of haematological disorders but blood tests are frequently used in a wide variety of other disorders. Why is this? Physicians do not often take biopsy samples from internal organs such as the liver or kidneys. Such biopsies involve trauma for the patient, are expensive for the hospital service, and the inevitable hospitalisation disrupts the life of the patient and the family. On the other hand, blood is an easily accessible tissue which provides a 'window' on many aspects of the body's metabolism.

Blood consists of a variety of cells suspended in an extracellular fluid, plasma. Sometimes the clinician is interested in the blood cells themselves. Their types, numbers and morphology may provide clues about their production in the bone marrow or the processes (and organs) responsible for their destruction. Their metabolism may be disturbed in a huge variety of conditions. Some which you will have heard of include leukaemia, anaemia, malaria, thalassaemia and glucose 6-phosphate dehydrogenase deficiency (obviously this is only a start!).

Blood cells may also be representative of other less accessible cell types. All nucleated somatic cells, including white blood cells, have the same genetic potential, i.e. they contain the complete genome even though they only express a small number of their genes. Ever since this important finding was made, DNA isolated from lymphocytes has been used in screening for genetic disorders which may be expressed in only a limited number of less amenable cell types. Blood cells may also express, at low levels, enzymes whose activities are more usually associated with other organs. This frequently allows the clinician to diagnose metabolic disorders without the need for a biopsy of the tissue principally affected.

On other occasions attention may be focused on the plasma. The clinician may be concerned about clotting factors, hormones, transport proteins, cholesterol levels, etc. The composition of plasma may also give clues about the internal environment of cells. Thus changes in the plasma concentrations of key metabolites, enzymes, electrolytes and many other constituents can often give valuable information about tissue damage, genetic disorders, nutritional deficiencies, and a multiplicity of other conditions.

The Problem

Two patients visited their doctors with symptoms of excessive tiredness, depression and loss of appetite. Jane D. was a 28-year-old vegetarian. Simon K. was a 26-year-old foundry worker.

Results of haematological investigations are in Table 6.1. In both cases the peripheral blood film showed moderate microcytosis, hypochromia, anisocytosis and poikilocytosis (Figure 6.1). In addition Simon's blood contained 4% reticulocytes (normal 0.5–2.5%). As a result, further tests were ordered (Table 6.2). A bone marrow sample from each patient was stained for iron (Figure 6.2).

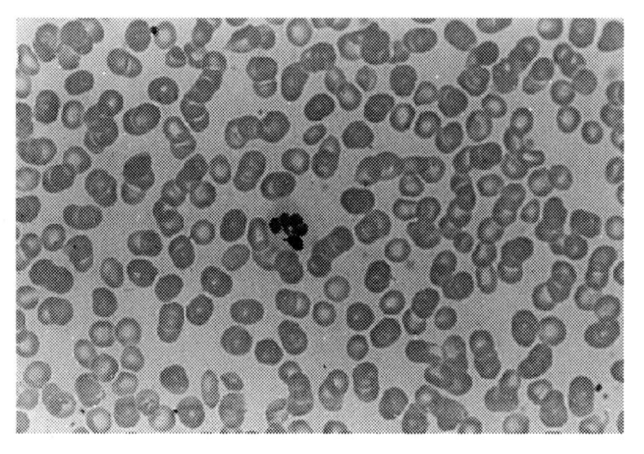

(a)

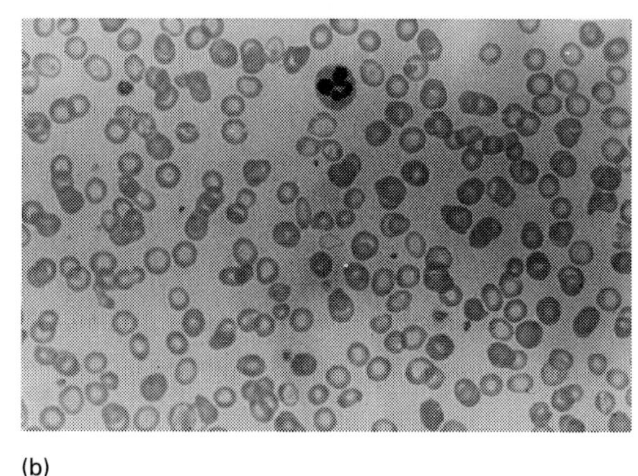

(b)

Figure 6.1. Peripheral blood films from normal subject (a) and Jane D. (b); a lymphocyte (nucleated cell) is present in each film for size comparison. Photographs kindly supplied by Ann Urmston, Manchester Royal Infirmary.

Table 6.1. Haematological investigations of the patients. M = male; F = female.

	Jane D.	Simon K.	Normal
Haemoglobin (g/dl)	9.0	10.2	13.0–17.0 (M) 11.0–15.0 (F)
Red cell count (per litre)	4.2×10^{12}	5.0×10^{12}	$4.5–6.5 \times 10^{12}$ (M) $3.9–5.6 \times 10^{12}$ (F)
Packed cell volume (PCV) (litre/litre)	0.3	0.36	0.41–0.53 (M) 0.35–0.48 (F)

Table 6.2. Further haematological tests of the patients.

	Jane D.	Simon K.	Normal
Plasma iron (μM)	7	38	10–30
Total iron-binding capacity (TIBC) of plasma (μM)	69	60	40–70

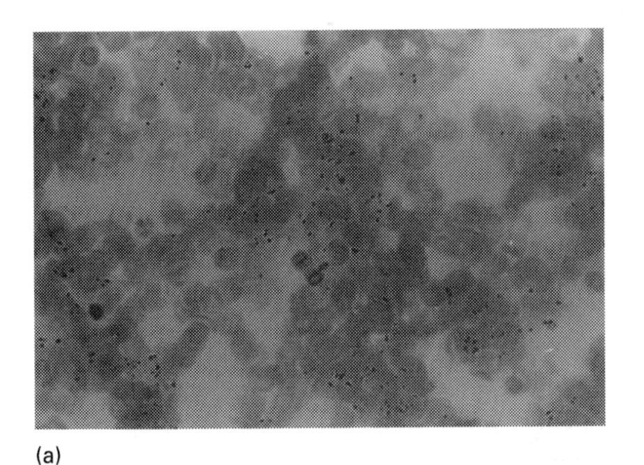

(a)

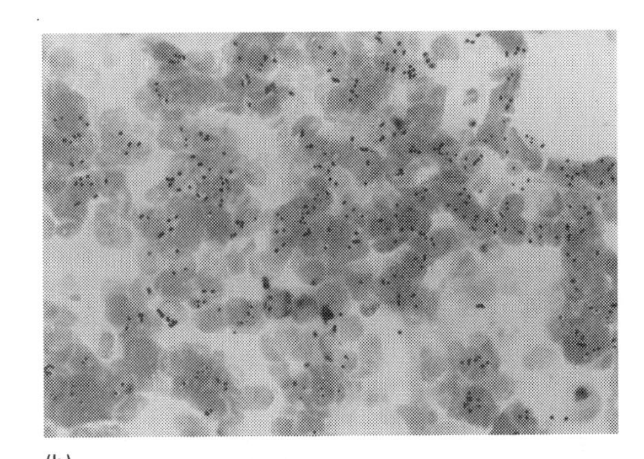

(b)

Figure 6.2. Bone marrow samples from a normal subject (panel a), Simon K. (panel b) and Jane D. (panel c) stained for iron stores. Iron stores are revealed as foci of dark staining against the tissue background. Photographs kindly supplied by Ann Urmston, Manchester Royal Infirmary.

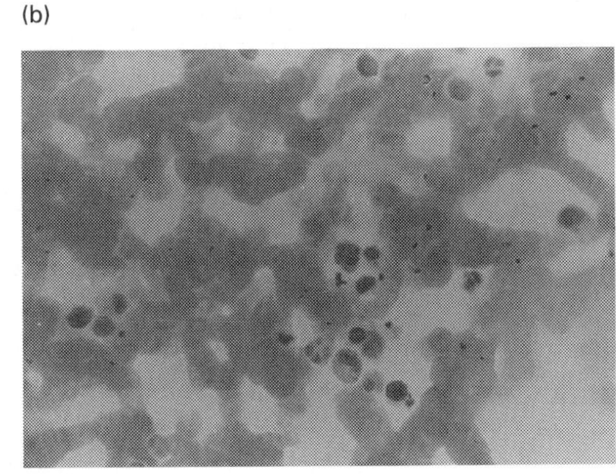

(c)

Questions

1. How would you measure the haemoglobin content of a blood sample?
2. From the findings in Table 6.1, what general condition affects both patients?
3. For each patient calculate the mean corpuscular volume (MCV) in fl (femtolitre = 10^{-15} litre), the mean corpuscular haemoglobin (MCH) in pg/cell (picogram = 10^{-12} g) and the mean corpuscular haemoglobin concentration (MCHC) in g/dl (decilitre = 10^{-1} litre). Normal values for these are 76–96 fl (MCV), 26–33 pg/cell (MCH) and 31–35 g/dl (MCHC).
4. How is iron carried in the plasma and stored in cells?
5. From the data in Table 6.2 what is the percentage saturation of the total iron-binding capacity in each patient?
6. What are the differences between the patients in the states of their iron stores and circulating iron?
7. Account for the decreased haemoglobin in each patient. Is there any information in their histories that might explain the observations?
8. How would you treat Jane D.?

Commentary

Haemoglobin is quantified by lysing the erythrocytes in hypotonic buffer or with detergent and measuring the haemoglobin by its characteristic absorption using a spectrophotometer. A comparison must be made against a standard solution of haemoglobin. However, different forms of haemoglobin present in blood have somewhat different absorption spectra so they are normally all converted into cyanmethaemoglobin.

Both patients are suffering from anaemia as indicated by the decreased haemoglobin levels. From the data in Table 6.1, the MCV, MCHC and MCH can be calculated as in Table 6.3. Note that in this problem the PCV is given as litre/litre which makes the calculations somewhat easier; however PCV is often expressed as a percentage (= l/l × 100).

Thus each patient has a red cell count within the normal range, but the erythrocytes are smaller than normal (microcytosis) and have a lowered haemoglobin content (MCH and MCHC), hence the hypochromia. This microcytic anaemia contrasts with other types of anaemia where the lower haemoglobin content of the blood may be due to a decrease in the number of erythrocytes.

Iron is carried in plasma bound to an iron carrier protein, transferrin. Transferrin binds to transferrin receptors on the surface of cells and undergoes endocytosis along with its iron. Inside the cell, the iron is stored bound to another protein, ferritin.

The plasma iron concentration indicates the amount of iron actually being carried by transferrin whereas the TIBC measures the total transferrin available. The percentage saturation of TIBC is calculated from the ratio of plasma iron : TIBC. The ranges for normal individuals are 25–55% (males), 15–50% (females). So for Jane D. it is very low (10%) and in Simon K. it is very high (63%). The bone marrow specimens, stained for iron (Figure 6.2), show Jane's iron stores are virtually depleted whereas Simon is overloaded with iron. All these data indicate that Jane has iron deficiency anaemia. Paradoxically, Simon is anaemic despite his saturated iron stores.

Jane's anaemia is readily accounted for by a shortage of iron to complete the synthesis of the haem required for haemoglobin. The enzyme ferrochelatase catalyses the incorporation of ferrous iron into protoporphyrin IX to form haem (Figure 6.3). Since the synthesis of globin chains is closely coupled to the availability of haem, haemoglobin synthesis is inadequate and the erythrocytes released from the bone marrow are small and deficient in haemoglobin. Iron deficiency is more likely to

Table 6.3. Calculation of MCV, MCH and MCHC.

	Calculation	Jane D.	Simon K.	Normal
MCV (fl)	$\frac{\text{PCV(l/l)}}{\text{red cells/l}}$	71	72	76–96
MCH (pg/l)	$\frac{\text{Hb(g/dl)}}{\text{red cells/l}}$	21	20	26–33
MCHC (g/dl)	$\frac{\text{Hb(g/dl)}}{\text{PCV(l/l)}}$	30	28	31–35

occur in women as a result of menstruation, pregnancy and lactation. Vegetarians, particularly vegans, are also at risk. In vegetarian diets, iron is less abundant and often poorly absorbed because vegetable fibre, phytates and tannins bind iron tightly. Typically only 1–5% of iron in vegetable sources is absorbed compared with 20–25% from red meat. Gastric HCl helps to release the iron, but for efficient absorption ferric iron, the usual dietary form, must be reduced to the ferrous state. In this respect, ascorbic acid (vitamin C) is an important dietary reductant. Jane should be treated initially with a course of ferrous sulphate and vitamin C tablets to see if her anaemia is reversed. Her diet should be examined to see if it is adequate with respect to iron and its absorption.

Iron deficiency is probably the world's most prevalent nutritional disorder, affecting about 60% of women and children in the Third World, of whom about one-third may be clinically anaemic as are perhaps 10–20% of women of child-bearing age in Western countries. Long-term iron deficiency is particularly serious in children, affecting brain development and leading to behavioural abnormalities.

In Simon K. the failure to synthesise sufficient haemoglobin cannot be ascribed to iron deficiency; rather there must be some block in iron utilisation. A clue to the cause is his occupation as a foundry worker. Here he may be exposed to toxic metals and this should be followed up. In fact lead poisoning would give rise to many of his symptoms, although this cannot be deduced directly from the data provided; a plasma lead assay would have to be done. Lead is extremely toxic, especially to the erythropoietic system, nervous tissue and the kidney. It has a particular avidity for the thiol groups of cysteine residues in proteins with the result that many enzymes which depend on thiol groups for their catalytic activity are irreversibly inhibited. Among these are two enzymes involved in the haem biosynthesis pathway (Figure 6.3). One is ferrochelatase, which catalyses the final step in the pathway, and the another is amino-lævulinic acid dehydratase which catalyses the formation of porphobilinogen from δ-aminolævulinic acid. Blocks in haem biosynthesis would lead to a build-up of

Figure 6.3. Pathway of haem biosynthesis.

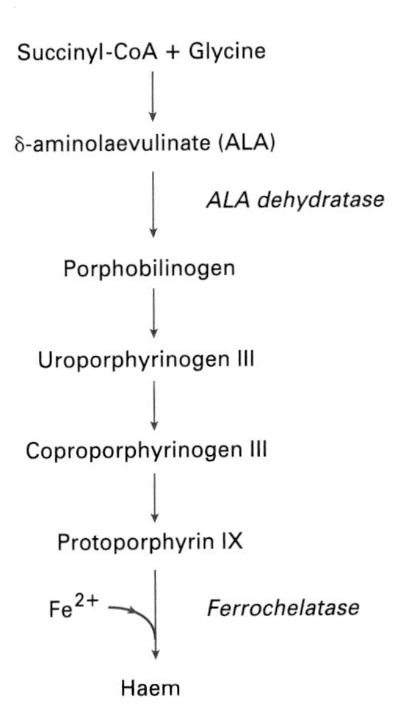

porphyrin intermediates and their metabolites. In lead intoxication, therefore, there is increased urinary excretion of coproporphyrin, δ-aminolævulinic acid and protoporphyrin, providing simple and sensitive indicators of lead ingestion. Because lead is deposited in the skeleton in exchange for calcium, its toxic effects persist over a long period unless skeletal lead is removed. Lead is a potent neurotoxin and children are particularly at risk. Damage to the kidneys would be manifest as glucosuria, proteinuria and amino aciduria. The usual sources of lead intoxication are lead-based paints, lead water pipes and the exhaust emissions from vehicles burning leaded petrol.

Further Questions

1. What are the dangers of continued treatment of Jane with dietary iron and how would you monitor her treatment with this in mind?
2. Suppose Jane failed to respond to dietary iron supplementation. What other causes for her iron deficiency would you consider and what measures would you take?
3. Why is treatment of iron-deficiency anaemia often accompanied by an early rise in the reticulocyte count?
4. How would you treat Simon K. to rid his body of ingested lead? What precautions would have to be taken in the treatment that you prescribe?
5. A block in the haem biosynthetic pathway results in overproduction of porphyrin intermediates prior to the block (Figure 6.3). Why is this?
6. Most cells contain proteins called metallothioneins which protect the cell against heavy metals such as lead, cadmium, mercury and zinc. Considering the information mentioned in this case about the way in which lead interacts with enzymes, how might metallothioneins combat the toxic effects of lead in cells?
7. Figure 6.1 shows how lymphocytes can be used as convenient size markers for comparison with erythrocytes. What types of white blood cells are normally present in blood and what are their origins and roles? Find out what the major leukaemias are and what their haematology would be like.
8. Lead irreversibly inhibits ferrochelatase. What is the difference between irreversible and reversible inhibition? What are the different types of reversible inhibition and how are they distinguished? Can you give examples of enzyme inhibitors used therapeutically? (See also Problems 16, 20 and 21.)

Connections

- Review your knowledge of the roles of erythrocytes, their life history and the histological picture presented in different pathological conditions. Check that you know how blood counts can be made, for instance using a haemocytometer or automated cell counter. From your Anatomy course, remember the uses of different histological stains used in haematology. (See also Problem 11.)

- Draw a diagram to illustrate the metabolism of iron in the body. This will include factors that control its absorption from the gut, transport in the blood, storage in tissues, synthesis of the major iron-containing proteins and their degradation. Which tissues will be involved in this complex story?
- This problem posed the question of how spectrophotometry could be used to quantify haemoglobin. Look up the absorption spectra of oxy- and deoxy-haemoglobins and devise a spectrophotometric method for measuring the extent of haemoglobin oxygenation. Check that you understand why the term *oxygenation* rather than *oxidation* is used.
- Erythropoietin is a hormone essential for erythropoiesis. Recently, recombinant erythropoietin has become available for treating patients. What patients would benefit? Check that you understand what *recombinant* means and how such human proteins are prepared. Find out what other recombinant human proteins are currently available for therapeutic use. What are the advantages and disadvantages of recombinant proteins?
- Blood also contains platelets, but do you remember what their role is in coagulation and inflammation? Also use this as an opportunity to review the process of blood clotting and the role of the complement system.
- This problem featured two forms of anaemia. Many haemoglobinopathies also give rise to anaemia. Check that you understand the origins of the anaemias associated with the sickle cell mutation and with thalassaemia. (See also Problem 11.)
- Lead interferes with normal bone formation. Review what you know about bone formation. This will include collagen biosynthesis and its post-translational modifications, calcium absorption, the roles of vitamin D, parathyroid hormone, calcitonin, growth hormone, phosphate metabolism, mineralisation, etc. (See also Problems 8 and 15.)
- Take this opportunity to review what you know about the major plasma proteins and their roles. How would you separate these various plasma proteins? Plasma enzymes are often used for diagnostic purposes: which ones and under what conditions? (See also Problems 3, 8, 11, 18 and 20.)

References

Balestra D J (1991) Adult chronic lead intoxication. *Archives of Internal Medicine* **151**, 1718–20.

Bottomley S S and Muller-Eberhard U (1988) Pathophysiology of heme synthesis. *Seminars in Hematology* **25**, 282–302.

Golde D W and Gasson J C (1988) Hormones that stimulate the growth of blood cells. *Scientific American* **259**(1), 34–42.

Pagliuca A and Mufti G J (1990) Lead poisoning; an age-old problem. *British Medical Journal* **300**, 830.

Ringenberg Q S, Doll D C, Patterson W P, Perry M C and Yarbro J W (1988) Hematologic effects of heavy metal poisoning. *Southern Medical Journal* **81**, 1132–9.

Scrimshaw N S (1991) Iron deficiency. *Scientific American* **265**(4), 24–30.

7 BOY OR GIRL?

Introduction

The human body comprises several thousand different cell types each with its own characteristic and specific roles to perform. Like any other well-ordered community of individuals, the cells of the body require efficient sensory and effector systems to detect perturbations in the external and internal environments and to bring about the appropriate metabolic responses. Successful coordination between these sensory and effector systems depends on well-organised intercellular communication. This is the role of the nervous and endocrine systems.

Endocrine organs in the body react to stimuli by secreting chemical messengers, the hormones, which travel via the bloodstream to elicit responses from target cells. Hormones are extremely heterogeneous chemically and include steroids (the subject of this problem), proteins and peptides such as insulin, catecholamines (for example adrenaline) and even the gas nitric oxide.

The steroid hormones comprise four main groups – the *glucocorticoids* such as cortisol and corticosterone which control carbohydrate metabolism and inflammation, the *mineralocorticoids* such as aldosterone which regulate electrolyte balance, *progestins* including progesterone, the hormone of pregnancy, and the *sex steroids* (principally oestradiol in women and testosterone in men) which are vital for sexual development, fertility and the secondary sexual characteristics. All of these are synthesised from cholesterol in a series of enzyme-catalysed steps in which progesterone occupies a central position in steroidogenesis (Figure 7.1). Normally the gonads are principally responsible for the production of the sex steroids, the corpus luteum for progesterone, and the adrenal cortex for the glucocorticoids and mineralocorticoids. However, as shown in Figure 7.1, the adrenal cortex is also able to produce small amounts of sex steroids and progesterone.

Steroids, both natural and synthetic, are used widely in medicine for reducing inflammation, suppressing the immune response in transplantation and controlling electrolyte balance. 'Anabolic' steroids are used, often illegally, in body-building. The first steroid antagonist used widely as a drug, the anti-mineralocorticoid spironolactone, is still used to control salt retention. Steroid antagonists are also of value in the treatment of breast and prostate cancers.

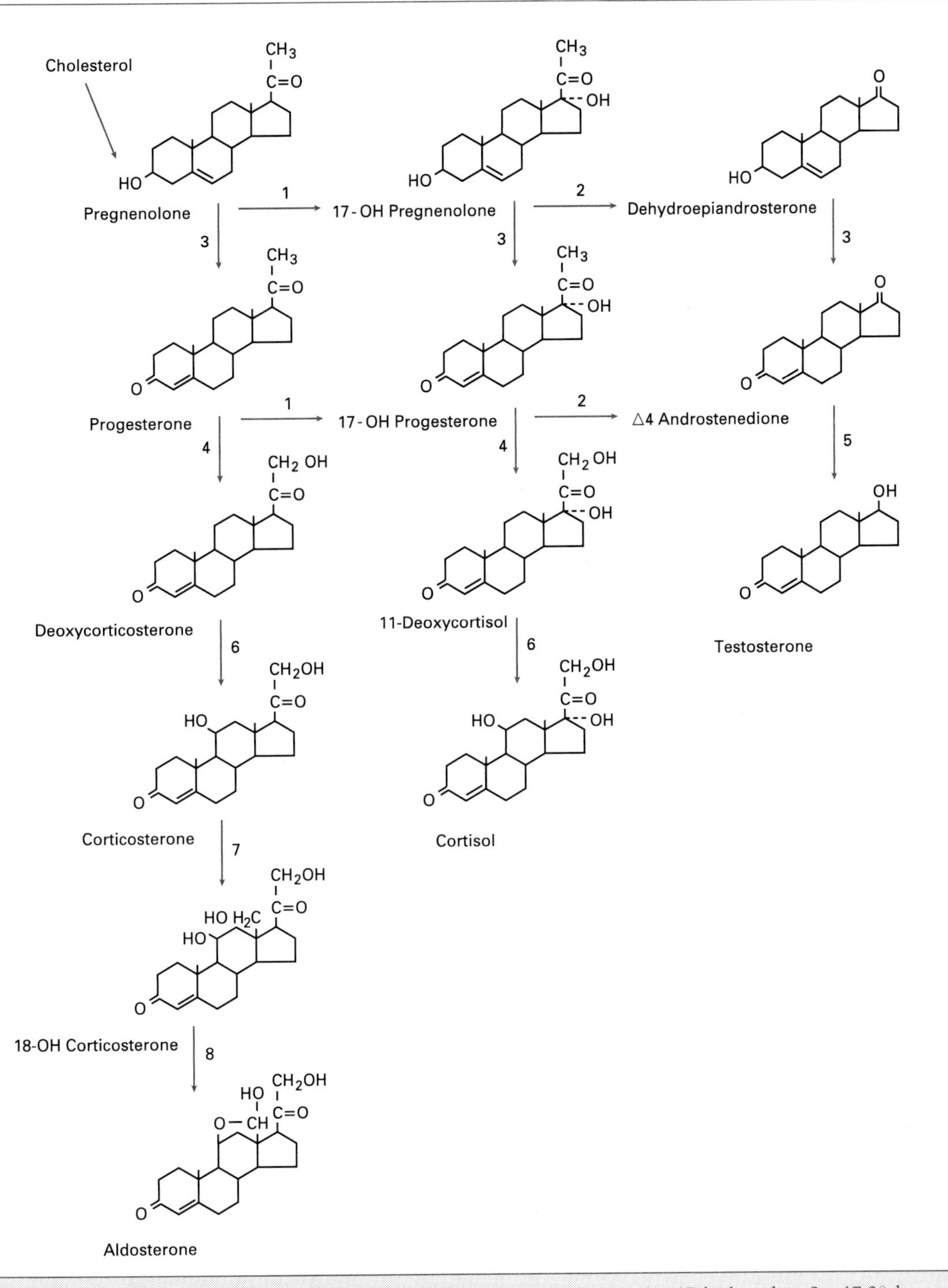

Figure 7.1. Pathways of adrenal steroid biosynthesis from cholesterol. *Key to enzymes*: 1 = 17-hydroxylase; 2 = 17,20-lyase; 3 = 3β-hydroxysteroid dehydrogenase; 4 = 21-hydroxylase; 5 = 17β-hydroxysteroid dehydrogenase; 6 = 11-hydroxylase; 7 = 18-hydroxylase; 8 = 18-hydroxysteroid dehydrogenase.

The Problem

A two-month-old male child, Andrew T., was admitted to hospital after failure to thrive. He had been born at home after a normal pregnancy. The midwife had decided the baby was 'probably a boy' though the external genitalia were ambiguous in appearance. Since then there had been a history of lethargy, refusing food and vomiting.

On admission he was found to be clinically dehydrated. Figure 7.2 shows the appearance of his external genitalia. The gonads could not be palpated. A blood sample was analysed (Table 7.1) and a metaphase chromosome spread was made from his peripheral lymphocytes (Figure 7.3).

Figure 7.2. Ambiguous external genitalia illustrative of the case in this problem. Photograph kindly supplied by Dr Maria New (New York Hospital – Cornell Medical Center, New York) and reproduced from New and Levine (1973) *Advances in Human Genetics* 4, 251–326, by permission of Plenum Press, New York.

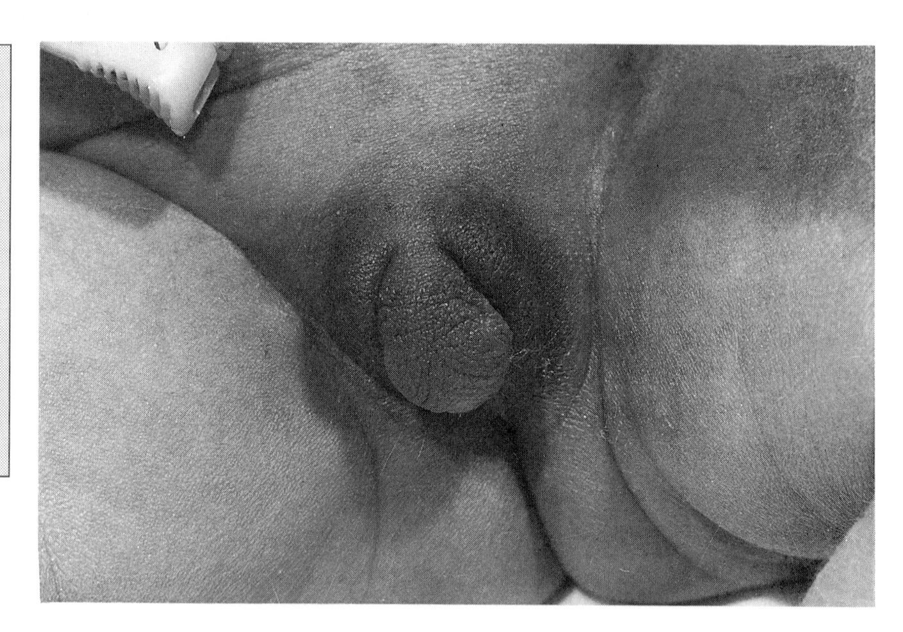

Figure 7.3. Metaphase chromosomes from Andrew's peripheral lymphocytes. Photograph kindly supplied by Jeff Williams, St James's University Hospital, Leeds.

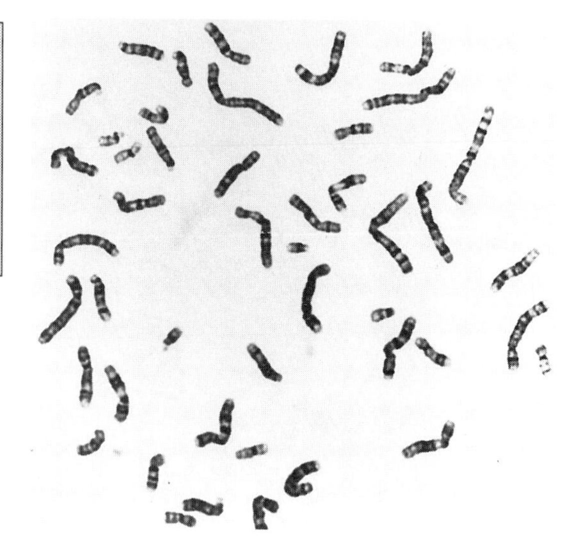

Table 7.1. Plasma analysis on admission. Apart from those steroids listed, there was no significant increase in the levels of other steroids.

	Patient	Normal controls
Urea (mM)	7.7	2.5–7.0
Glucose (mM)	3.9	3.5–5.5 (fasting)
Na^+ (mM)	122	134–147
K^+ (mM)	7.1	3.5–5.0
17-hydroxyprogesterone (nM)	670	<15
Cortisol (nM)	25	150–600 (at 9 a.m.)
Testosterone (nM)	8	0.5–3.2 (adult female) 5–40 (adult male)

Questions

1. Are any of Andrew's plasma constituents (Table 7.1) abnormal? Why was the cortisol measurement done at 9 a.m.?
2. Examine the pathways of steroid biosynthesis from cholesterol in the adrenal cortex outlined in Figure 7.1 and deduce the position of any metabolic block present in Andrew. What important steroids will be deficient?
3. Account for any electrolyte imbalance. Why is Andrew dehydrated?
4. What can you deduce about Andrew's karyotype? Hint: can you pair up all the chromosomes?
5. Why have his external genitalia developed abnormally?
6. What immediate intervention is necessary to preserve Andrew's life?
7. Andrew's long-term endocrinological treatment included cortisol replacement therapy. Within 3 weeks this resulted in his plasma 17-hydroxyprogesterone and testosterone levels falling to 60 nM and 1.5 nM, respectively. Why did these changes occur?
8. Why would treatment with cortisol alone be insufficient?
9. In this case adrenal hyperplasia occurs. What is hyperplasia? What hormone might be responsible?

Commentary

Cortisol production from the adrenal cortex shows a pronounced diurnal variation; its plasma levels are maximal shortly after waking and decline during the day to their low point in late evening. Consequently, it is important to use a standard time for cortisol measurements. Andrew's plasma cortisol level is low whereas his 17-hydroxyprogesterone and testosterone levels are clearly well above normal. He has hyponatraemia but hyperkalaemia; his plasma urea is slightly raised.

The elevated 17-hydroxyprogesterone and low cortisol, in the absence of any significant level of 11-hydroxysteroids, points to a deficiency of the steroid 21-hydroxylase enzyme (Figure 7.1). The levels of metabolites proximal to this enzyme step will be raised but distal to this steroids will be deficient, of which the most important are the glucocorticoids, cortisol and corticosterone, and the principal mineralocorticoid, aldosterone.

The mineralocorticoid deficiency has profound effects on water and electrolyte balance. The kidney produces about 170 litres per day of glomerular filtrate whose composition is virtually identical to plasma except for the absence of most plasma proteins. To form the 1–2 litres of urine normally voided each day, the composition of the glomerular filtrate must be modified during its passage through the kidney tubule. Aldosterone promotes Na^+ reabsorption and K^+ excretion in the distal tubule so Andrew's lack of aldosterone results in hyponatraemia and hyperkalaemia (Table 7.1). Na^+ is the most important ion contributing to the osmotic potential of the extracellular fluid, so Na^+ reabsorption in the kidney tubule normally also results in water reabsorption. Unfortunately, Andrew's Na^+ depletion ('salt wasting') leads to a fall in the osmotic potential of his extracellular fluid. Consequently his renal water reabsorption is low, leading to dehydration.

Although the patient is assumed to be male, there is no evidence of a Y chromosome (Figure 7.3). All the metaphase chromosomes are present as diploid pairs including the X chromosomes. This is easily seen when the metaphase chromosomes have been arranged as a conventional karyotype (Figure 7.4). Andrew's karyotype is 46 XX and he is really a genotypic female (Figure 7.4 also shows a normal male karyotype, 46 XY, for comparison). What seems to have happened is that the patient's external genitalia have been masculinised *in utero* and the midwife made an incorrect gender assignment. The virilisation is presumably the result of exposure of the developing genitalia to excessive concentrations of testosterone (see Table 7.1), the principal androgenic steroid hormone.

Why are excessive amounts of testosterone synthesised? The metabolic block in the patient leads to the loss of the normal negative feedback regulation of steroidogenesis which is exerted by cortisol via the anterior pituitary and adrenocorticotropin (ACTH). Consequently, large amounts of 17-hydroxyprogesterone build up, much of which is diverted towards the synthesis of testosterone. Absence of this feedback control on the pituitary via cortisol also results in overproduction of ACTH, which in turn stimulates excessive growth of the adrenal glands (hyperplasia). Hence this group of disorders is referred to as the *congenital adrenal hyperplasias*.

The immediate threat to life is the dehydration and Na^+ loss which calls for infusion of intravenous saline. Then hormone replacement therapy must be instituted and this should include both cortisol (usually given as cortisone) and a mineralocorticoid (e.g. 9α-fludrocortisone, a synthetic mineralocorticoid). This replacement therapy not only supplies the missing glucocorticoid, cortisol, but also

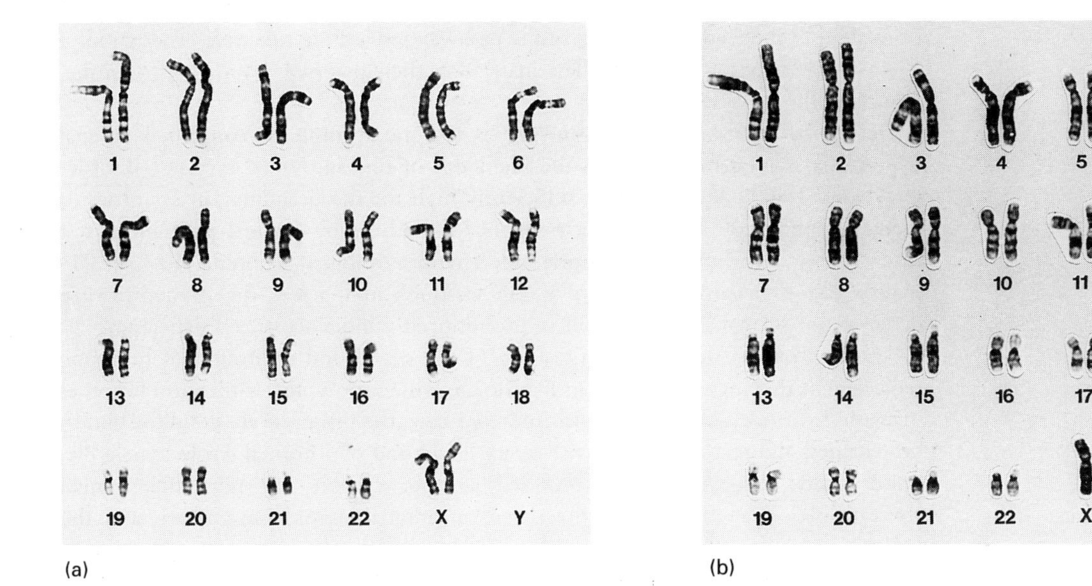

Figure 7.4. Karyotype of the patient (panel a) and a normal male (panel b). In panel a, the chromosomes shown in Figure 7.3 have been arranged in diploid pairs. Photographs kindly supplied by Jeff Williams, St James's University Hospital, Leeds.

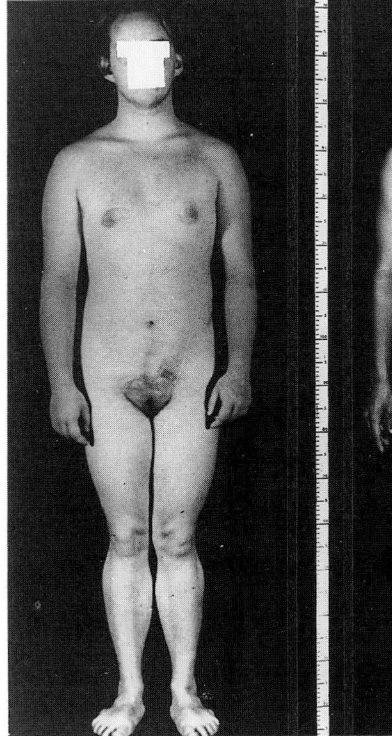

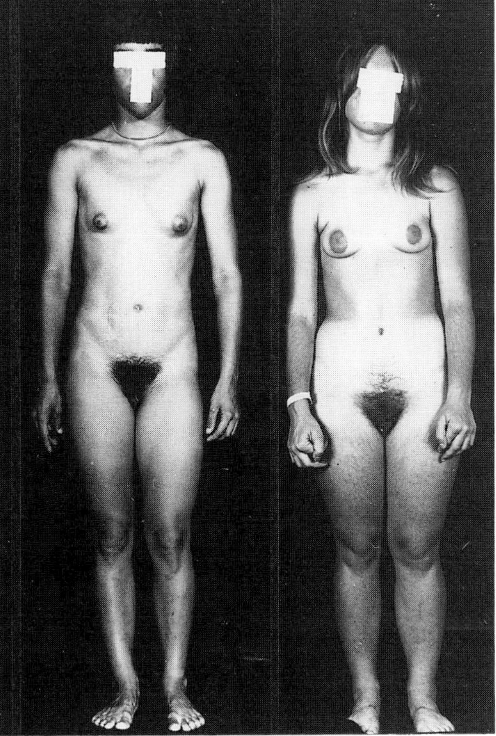

Figure 7.5. Habitus of pubertal girls with congenital adrenal hyperplasia due to 21-hydroxylase deficiency. The patients were treated (left to right) from age 16 years, age 9 years and age 4 years. Photograph kindly supplied by Dr Maria New (New York Hospital – Cornell Medical Center, New York) and reproduced from New and Levine (1973) *Advances in Human Genetics* 4, 251–326, by permission of Plenum Press, New York.

re-establishes the feedback control of the pathway preventing excessive production of 17-hydroxyprogesterone. Much less material is then diverted towards testosterone, removing the virilisation stimulus.

Deficiency of steroid 21-hydroxylase is the most common congenital adrenal hyperplasia. In northern Europe, the incidence of this autosomal recessive disorder may reach 1:5000 (Appendix 4), sufficiently high for the Scandinavian countries to include it among the genetic defects screened for at birth by the heel-prick blood test (the others being phenylketonuria and hypothyroidism). About 75% of 21-hydroxylase deficiency cases involve 'salt wasting'; in the rest, the 21-hydroxylase defect is not so marked as to result in pronounced mineralocorticoid deficiency.

The importance of early identification of the defect and institution of hormone replacement therapy is illustrated in Figure 7.5, which shows that virilisation becomes increasingly irreversible with age. In Andrew's case, the prospects are good for him to be renamed and reared as a girl with every likelihood of a normal fertile female life. Some plastic surgery may be required later to correct any remaining genital abnormalities. The location of the defect in adrenal steroid biosynthesis and the virilisation explains the alternative name for this disorder, the *adrenogenital syndrome.*

Further Questions

1. In what ways would the symptoms of 21-hydroxylase deficiency have been different in a chromosomal male?

2. What would the outward signs of dehydration be in a young child?

3. In this condition plasma ACTH levels will be elevated. How would you assay ACTH?

4. Chronic treatment with large doses of glucocorticoids is used to treat certain conditions. Do you know of any situations where this type of treatment would be used? What effects will this chronic exposure to glucocorticoids have on a patient's pituitary and adrenal function? What is likely to occur if the treatment is abruptly withdrawn?

5. What diagnostic feature of 21-hydroxylase deficiency would be suitable for the neonatal screening programme mentioned in the *Commentary*?

6. Another type of congenital adrenal hyperplasia, 11β-hydroxylase deficiency, also results in virilisation, but has a markedly different effect on electrolyte balance. In what way would electrolyte balance be disturbed and how is this brought about? Why may 11β-hydroxylase deficiency also be characterised by hypertension? (See also Problem 16.)

7. To obtain metaphase chromosomes suitable for karyotyping, cells are treated with a drug to arrest them in mitosis. What type of drug would be used and what would be its subcellular target?

8. Find out about the anabolic steroids and the non-steroidal oestrogen, diethylstilboestrol. What do they do, how is their use detected and what are their dangers?

Connections

- Make a list of all the hormones with which you are familiar and check that you understand where they are synthesised, what their targets are and how they act. This should make you realise that the way in which steroid and thyroid hormones elicit their responses is quite different from that of other hormones (such as the polypeptide hormones and adrenaline) and focus your mind on receptors and second messengers. (See also Problems 12, 15 and 22.)
- Review the areas of electrolyte homeostasis, mineral metabolism and regulation of blood pressure. This should recall for you the renin/angiotensin system, parathyroid hormone, calcitonin, vitamin D, growth hormone, atrial natriuretic hormone, the endothelins, nitric oxide, the transport of ions, glucose and water. Can you think of others? (See also Problems 15 and 16.)
- Review the process of sexual development and the roles of the X and Y chromosomes. Ensure you understand the terms 'genotype', 'phenotype', 'diploid', 'haploid', 'dominant', 'recessive'. Revise the structure of chromosomes and the processes of DNA replication, mitosis and meiosis. (See also Problem 22.)
- Karyotyping was part of the diagnosis in this condition. Remember that other conditions can be diagnosed by this technique, such as Down's syndrome and certain cancerous and pre-cancerous conditions. Check that you are familiar with these and other similar conditions.
- The problem mentioned the heel-prick test for phenylketonuria. Check you know what this disorder is and how the test works. Use this as an opportunity to review pre-natal and neonatal screening methods for genetic disorders. (See also Problem 19.)
- The heel-prick test is also used for diagnosing hypothyroidism. Find out about this disorder and use it to recall the nature of the thyroid hormones, their synthesis, regulation and actions.

References

Hughes I A (1990) Congenital adrenal hyperplasia. *Trends in Endocrinology and Metabolism* 1, 123–8.

Pang S and Clark A (1990) Newborn screening, prenatal diagnosis and prenatal treatment of congenital adrenal hyperplasia due to 21-hydroxylase deficiency. *Trends in Endocrinology and Metabolism* 1, 300–7.

Strachan T (1989) Molecular genetics of congenital adrenal hyperplasia. *Trends in Endocrinology and Metabolism* 1, 68–72.

8 BRITTLE BONES OR BATTERED BABY?

Introduction

The circus 'indiarubber man', with his hyperextensible skin and joints, is a classic example of a variety of human diseases in which there are inherited defects in the structure of particular connective tissue proteins. The famous virtuoso violinist, Paganini, may also have 'suffered' from such a disorder, which could well explain his long thin limbs (dolichostenomelia), spider-like fingers (arachnodactyly) and hence his unique musical talents. In Paganini's case, the inherited defect was probably Marfan's syndrome which is estimated to affect 1 in 10 000 people (Appendix 4). President Abraham Lincoln's physique has also led to the suggestion that he too suffered from Marfan's syndrome which has recently been identified as due to a defect in the connective tissue protein, fibrillin. There has even been a proposal to test Lincoln's DNA, isolated from museum specimens, for fibrillin gene mutations. Some of today's basketball 'stars' may also owe their prowess to Marfan's syndrome. The main life-threatening complications of such disorders are cardiovascular in nature, especially aortic aneurysms, although the likelihood of aortic rupture can be minimised by treatment with β-adrenergic blocking drugs or by surgery.

The best understood connective tissue disorders are those affecting collagen. Collagen is an extracellular protein synthesised by, and secreted from, a variety of cells (such as fibroblasts) and organised into insoluble fibres of high tensile strength. Along with the proteoglycans, collagen is a major component of the extracellular matrix which surrounds animal cells and provides mechanical strength and rigidity to tissues and organs. There are at least five major types of collagen that occur in different tissues. Each collagen has its own distinctive features, although all are composed of a characteristic triple helical structure. Collagen has a very unusual amino acid composition with a high content of glycine and proline and the presence of the modified amino acids hydroxyproline and hydroxylysine. The biosynthesis of collagen provides an excellent example of the post-translational modification of a protein involving hydroxylation, glycosylation, proteolytic processing and cross-linking reactions. Defects in any of these essential processes can lead to structural abnormalities and consequently disease states. The present problem is one such example.

The Problem

A female child, Susan M., had been born after a normal pregnancy. She had a history of frequent long-bone fractures resulting from only minor injury and necessitating treatment at her local hospital emergency room. Child abuse had initially been suspected but as she developed, skeletal abnormalities became apparent which were confirmed by radiography. At 3 years of age she could still only crawl and had difficulty standing.

Susan's type I collagen, the principal collagen of bone, was investigated. Fibroblasts were cultured from a skin biopsy and the type I collagen which they synthesised was analysed by polyacrylamide gel electrophoresis in the presence of sodium dodecylsulphate (SDS–PAGE). Type I collagen consists of two $\alpha1$(I) polypeptide chains and one $\alpha2$(I) chain. Some of the patient's $\alpha1$(I) polypeptide chains had an unusually low electrophoretic mobility in the absence of 2-mercaptoethanol, but co-migrated with normal $\alpha1$(I) chains in the presence of this thiol reagent (Figure 8.1).

The infant was shown to be heterozygous for a mutation in the COL1A1 gene which encodes the procollagen precursor of $\alpha1$(I) collagen. A comparison of part of the nucleotide sequences (the coding, or non-transcribed strand) of the normal and mutant COL1A1 genes is shown in Figure 8.2.

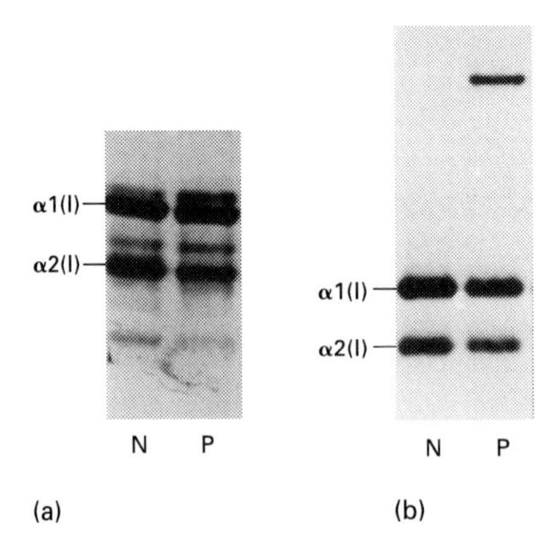

Figure 8.1. SDS–PAGE of collagen polypeptide chains. Collagens from the patient (P) and a normal person (N) were separated in the presence (panel a) and absence (panel b) of 2-mercaptoethanol. Note the slowly-migrating $\alpha1$(I) collagen (←) in the patient's sample. Adapted from an original figure kindly supplied by Professor Peter Byers (University of Washington, Seattle).

Normal.....GCCTCGGGGTGCCAATGGTGCGCCT**G**GCAACGACGGTGCTAAGGGCGATGCCGGT...

Patient....GCCTCGGGGTGCCAATGGTGCGCCT**T**GCAACGACGGTGCTAAGGGCGATGCCGGT...

Figure 8.2. Comparison of the nucleotide sequences for the COL1A1 genes in the patient and a normal subject. Only part of the gene sequence is shown; the bases involved in the mutation are shown in **bold**.

Questions

1. Use the Genetic Code (Appendix 6) to predict the effect of the COL1A1 mutation. Is this supported by the electrophoretic data in Figure 8.1?
2. Why are only some of the $\alpha 1$(I) collagen chains affected?
3. What are the biochemical consequences of this change likely to be on the assembly of type I collagen?
4. How might the predicted changes result in the skeletal deformities and the brittle bones?
5. On further investigation the disorder was shown to have a dominant pattern of inheritance in the patient's family. How can this observation be explained? (Hint: consider the structure and assembly of type I collagen.)
6. Susan's mother is pregnant again. How you would determine whether or not the foetus is likely to suffer from the same defect?

Commentary

Although child abuse may be involved in cases of unexplained bone fractures, abnormal bone development should also be considered. The condition described here is one example of a number of genetic disorders known as *osteogenesis imperfecta* which are often accompanied by other features including blue sclerae, abnormal teeth (dentinogenesis imperfecta), thin skin and progressive hearing loss. In most instances, such as the one described here, osteogenesis imperfecta is caused by a point mutation in one of the genes encoding the chains of type I procollagen which results in the substitution of a glycine residue for another amino acid. The resultant abnormal protein chain is incorporated into the triple helix of collagen, rendering it structurally and functionally defective.

By convention, gene sequences are recorded in data banks in the form of the nucleotide sequence of the sense (coding) strand, equivalent to the mRNA except that the pyrimidine base thymine (T) in the gene would be replaced by uracil (U) in mRNA. The COL1A1 gene sequence in Figure 8.2 is theoretically capable of being read (conventionally from left to right) in triplet codons in three different reading frames. In practice, the ribosomes read the mRNA in the correct frame starting at the initiating codon. Although the reading frame is not specified in Figure 8.2, only one of the three possibilities encodes an amino acid sequence with the characteristics of collagen, i.e. with glycine every third residue. Glycine is essential for the formation of

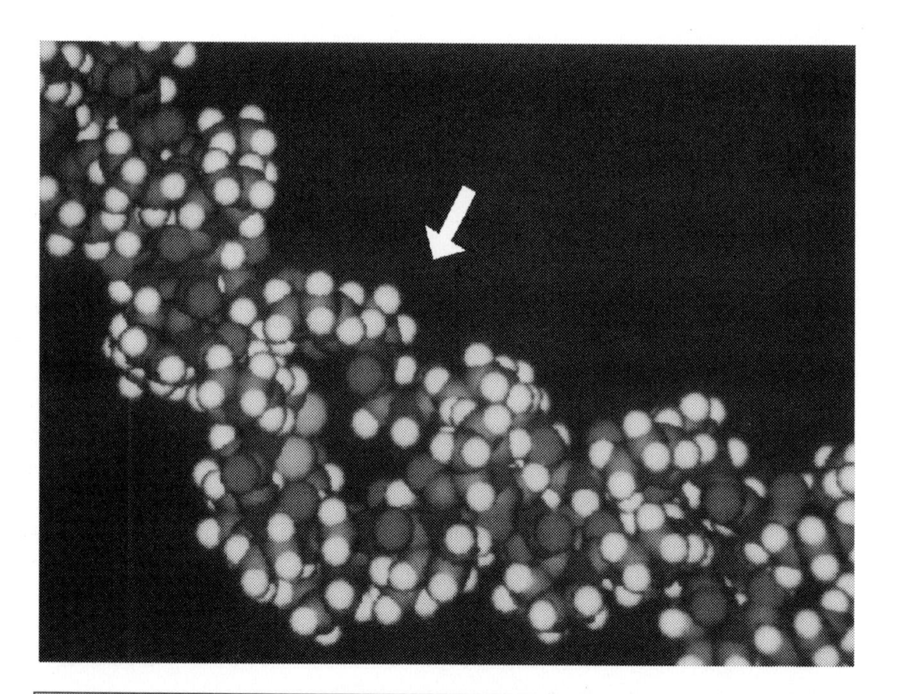

Figure 8.3. Space-filling model of mutated collagen predicting the 'distortion' in the collagen triple helix due to a cysteine/glycine substitution (←). Photograph kindly supplied by Dr Darwin Prockop (Jefferson Medical College, Philadelphia) and reproduced from Vogel *et al.* (1988) *Journal of Biological Chemistry* 263, 19 249–55, by permission of the American Society for Biological Chemistry and Molecular Biology.

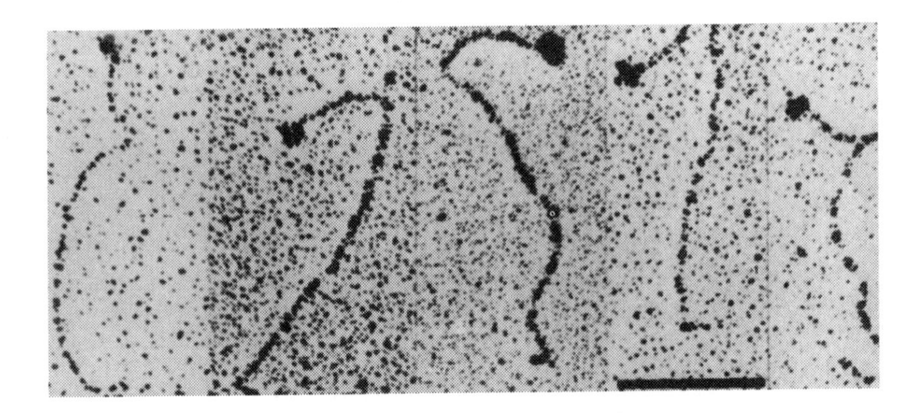

Figure 8.4. Electron micrographs of rotary shadowed type I procollagen molecules. A 'kink' can be seen in the triple helix domain at position 748 where there is a glycine → cysteine substitution. The globular 'head' is the procollagen C-terminal peptide. Bar = 100 nm. Photograph kindly supplied by Dr Darwin Prockop (Jefferson Medical College, Philadelphia) and reproduced from Vogel *et al.* (1988), reference as Figure 8.3.

the collagen triple helix since it is the only amino acid possessing a side chain (H–) small enough to be accommodated on the inside of the helix.

In the patient, the single base mutation (G → T) changes a glycine codon into one for cysteine. Cysteine possesses a bulkier side chain than glycine, distorting the collagen triple helix (Figure 8.3). The distortion may be sufficient to introduce a pronounced 'kink' into the procollagen chains, which can be detected using the electron microscope (Figure 8.4). The cysteine side chain also contains a reactive sulphydryl group (–SH). Not only is the formation of the collagen triple helix disrupted but there is also inappropriate disulphide cross-linking between the two $\alpha 1(I)$ collagen chains in the triple helix. This is revealed by SDS–PAGE (Figure 8.1) in which the denatured polypeptide chains migrate on the basis of their relative molecular masses (M_r). Any polypeptide chains that are cross-linked together via disulphide bonds (S–S) will migrate much more slowly than the individual chains. However, the sulphydryl (thiol) reagent, 2-mercaptoethanol, will cleave interchain disulphide bonds, allowing the constituent polypeptide chains to migrate according to their M_r.

Because the child is heterozygous for the mutation, only some of her $\alpha 1(I)$ chains behave anomalously in SDS–PAGE. Since she has one normal and one mutant allele, it might be expected that 50% of the chains would be abnormal but this assumes that the mutated gene is expressed at the same rate as the normal gene and that the abnormal protein is stable, neither of which appears to be the case in many genetic disorders.

Bone contains predominantly type I collagen which acts as a protein matrix for the deposition of hydroxyapatite (a form of calcium phosphate). The abnormal collagen structure in the patient results in defective mineralisation, skeletal abnormalities and poor mechanical strength. This form of osteogenesis imperfecta is a dominant negative disorder in that the heterozygote is affected as well as the homozygote. During

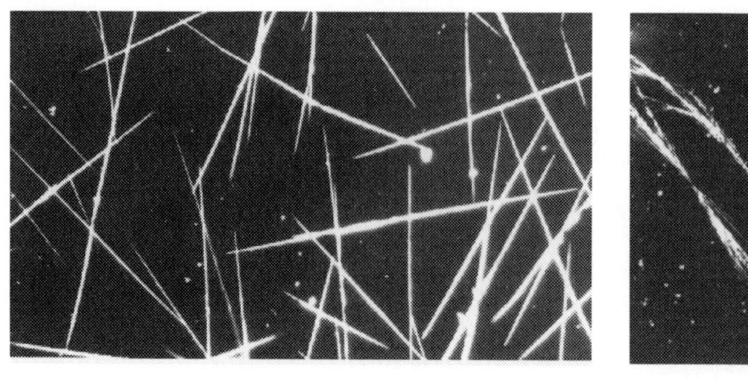

(a)

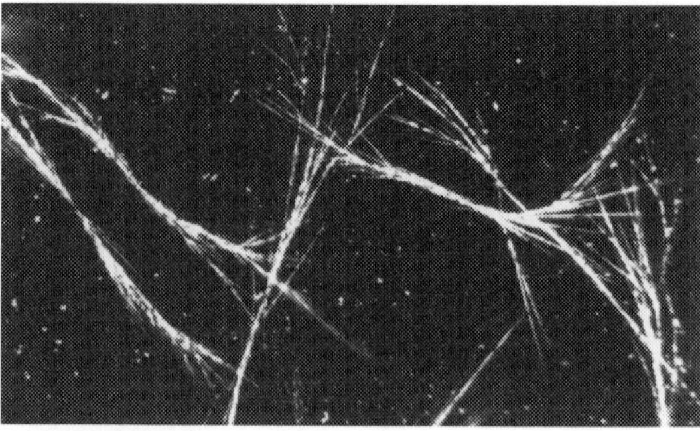

(b)

Figure 8.5. Light micrographs of collagen fibrils taken using dark-field illumination. Normal type I collagen (a) and a mixture of 90% normal and 10% cysteine-substituted collagen (b). Photograph kindly supplied by Dr Darwin Prockop (Jefferson Medical College, Philadelphia) and reproduced from Prockop *et al.* (1989), *Biophysics* 3, 81–9, by permission of Pergamon Press.

collagen assembly two $\alpha1$(I) chains are required together with one $\alpha2$(I) chain to form the triple helix. Assuming that both normal and defective $\alpha1$(I) chains are produced in approximately equal amounts by the fibroblast, one can predict that 75% of the type I collagen assembled will contain a defective chain and thus the heterozygote is unlikely to escape the deleterious consequences of the mutation. A nice illustration of the devastating consequences of such a mutation on the assembly of collagen is given in Figure 8.5. Here normal type I collagen fibrils are compared with those formed from a mixture of normal collagen and cysteine-substituted collagen.

It would be impractical and risky to try and detect the genetic abnormality by analysing foetal collagens. Genetic screening is safer as well as more convenient,

Figure 8.6. Use of allele-specific oligonucleotides (ASOs) to screen for the COL1A1 defect.

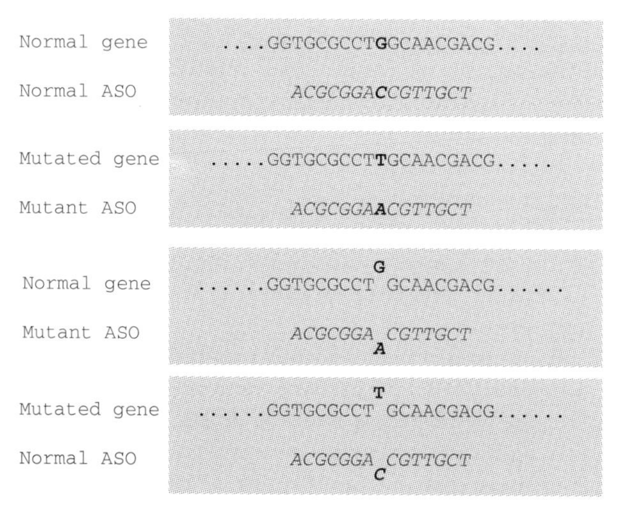

accurate and sensitive; it can also be done at a much earlier stage in gestation. Foetal DNA must first be obtained, usually from a small sample of chorionic villus (the foetal portion of the placenta). The DNA is then analysed in a blotting procedure using a pair of allele-specific oligonucleotide probes (ASOs). These are base-complementary to the short region of the COL1A1 gene spanning the mutational site, but differ from one another by the single base corresponding to the mutation (Figure 8.6). Under highly stringent hybridisation conditions (determined by the temperature and/or salt concentration), the normal ASO will hybridise only to the normal gene sequence and vice versa for the 'mutant' ASO.

Further Questions

1. Find out what you can about the general structural features of collagen genes and check your knowledge of the structure and the assembly of collagen itself. Why are mutations in collagen structural genes potentially much more damaging than those in many other genes?

2. Genetic defects are also known that affect type II collagen, chain composition $\alpha(II)_3$. Why are heterozygotes who express the defective gene much more seriously affected than those where the mutation 'silences' the gene?

3. What do you predict will be the biochemical consequences of mutations affecting certain lysine residues in type I collagen?

4. The consequences of a mutation replacing a glycine residue in collagen depend on the precise position of the glycine in the amino acid sequence. Generally the closer the replacement is to the *C*-terminus, the more serious the resulting phenotypic effects. What does this suggest concerning the assembly of the collagen triple helix?

5. Find out about the Ehlers–Danlos syndromes, which comprise several different disorders of collagen synthesis and assembly. What are the biochemical defects and clinical consequences of some of these?

6. What are the effects on collagen of dietary deficiencies in vitamin C (ascorbic acid) or of copper?

7. How is collagen degraded? Suggest why collagen degradation may be important in cancer and in infections involving *Clostridia.*

Connections

- Use this problem to review the general background of the structure of bone, mineralisation, cartilage, extracellular matrix, osteoblasts and osteoclasts, and their controlling factors: calcium, calcitonin, parathyroid hormone, vitamin D. (See also Problem 15.)

- The biochemistry of collagen should remind you of a variety of topic areas including protein structure, through post-translational modifications, to its major role in the extracellular matrix. List the various types of collagen and give an account of the classification system as well as the tissue location of the different types.
- Check that you understand the major structural features of proteins including the α-helix and β-sheet, and the types of bonds involved, such as hydrogen bonds, hydrophobic interactions, ionic interactions and disulphide bridges. Distinguish between an α-helix and the triple helix found in collagen.
- Consider the different possible types of mutations, dominant and recessive inheritance, and the consequences when a mutated polypeptide forms part of an oligomeric structure. Think about how mutations may be detected, especially pre-natally, and what ethical considerations this raises. (See also Problems 19 and 20.)
- The use of SDS–PAGE in this problem connects with separation techniques in general. Ensure you understand the basis of SDS–PAGE and the significance of whether a disulphide reducing reagent is present. Summarise what other types of electrophoresis you know and the basis of the separation. Would SDS–PAGE be appropriate if you were trying to separate enzyme activities in a serum sample? (See also Problems 3, 11 and 18–20.)
- Make sure you understand how hybridisation with a DNA probe can detect a mutation and why the conditions for hybridisation are crucial. (See also Problems 19 and 20.)

References

Byers P H (1990) Brittle bones – fragile molecules: disorders of collagen gene structure and expression. *Trends in Genetics* 6, 293–300.

Hulmes D J S (1992) The collagen superfamily – diverse structures and assemblies. *Essays in Biochemistry* 27, 49–67.

McKusick V A (1991) The defect in the Marfan syndrome. *Nature* **352**, 279–81. *News and views* article.

Paterson C R and McAllion S J (1989) Osteogenesis imperfecta in the differential diagnosis of child abuse. *British Medical Journal* **299**, 1451–54.

Prockop D J (1990) Mutations that alter the primary structure of type I collagen. *Journal of Biological Chemistry* **265**, 15 349–52. Mini-review.

Sykes B (1990) Bone disease cracks genetics. *Nature* **348**, 18-20. *News and views* article.

Weatherall D J (1991) *The New Genetics and Clinical Practice*, 3rd edn, Oxford University Press; particularly Chapter 4 The techniques of gene analysis, Chapter 9 Carrier detection and prenatal diagnosis of genetic disease and pp 158–61 on collagen disorders.

9 INCORRECTLY ADDRESSED?

Introduction

Ultrastructural examination of cells shows that they contain a large number of organelles bounded by membranes based on the lipid bilayer motif. The cell itself is separated from its external environment by another membrane, the plasmalemma. In addition an extensive membrane system, the endoplasmic reticulum (ER), permeates more or less the whole of the cytoplasm, especially in cells which secrete large amounts of protein. The cell is thus compartmentalised and also separated from its immediate environment. These subcellular compartments and their membranes are distinguished by their own characteristic range of structural proteins and enzymes. Furthermore, some of the cell's proteins must be secreted through the plasmalemma to form part of the extracellular matrix or to enter the circulation. However, with few exceptions, synthesis of most of the cell's proteins takes place, or at least starts, in the cytosol, the 'soluble' part of the cytoplasm between the various organelles. Consequently, the cell is faced with a major logistical problem of ensuring that newly-synthesized proteins reach their correct destination, be that the interior of an organelle, one of the cell's many membrane systems, or the outside of the cell.

Clearly it is necessary for newly-synthesized proteins to have some sort of address 'label' or 'tag' on them and for the cell to have a sorting and delivery system akin to a postal service. Examples of cellular address tags are the 'pre-' sequences found on proteins destined to pass into the ER and eventually be secreted from the cell. The tag is a short hydrophobic sequence of amino acids encoded at the *N*-terminus of the protein. The tag is the first part of the protein synthesized by ribosomes in the cytosol but which signals the binding of those ribosomes to the ER. This is followed by the 'threading' of the nascent polypeptide chain through the ER membrane into the ER interior whence it passes via the Golgi apparatus and secretory vesicles to the exterior of the cell. This is just one example; to sort other proteins with different destinations a variety of 'post-codes' or 'zip-codes' is needed. Obviously, any defect in the post-code or failure of the 'sorting office' can have very serious consequences (Figure 9.1) as this problem illustrates.

Figure 9.1. Address unknown, unable to deliver.

The Problem

A baby girl, Jane P., showed abnormal mental and physical development. Examination showed severe and multiple skeletal deformity, corneal clouding, cardiomegaly and hepatomegaly. Fibroblasts, obtained from skin biopsies, and circulating monocytes contained numerous inclusion bodies in their cytoplasm (Figure 9.2). These membrane-bound vacuoles were rich in complex polysaccharides, proteins and lipids. Normal cells are devoid of these inclusion bodies.

Skin fibroblasts from the patient and normal control subjects were cultured and the cells and medium were analysed for a wide range of acid hydrolases. The results for a representative enzyme, β-glucuronidase, are shown in Table 9.1. These acid hydrolases are efficiently retained by normal cells although a small amount is secreted into the medium. This situation mirrors what happens *in vivo*; the activities of most acid hydrolases in Jane's plasma were in large excess (about 10-fold to 200-fold) over the mean normal activities.

When Jane's fibroblasts were incubated in medium that had previously been used to grow *normal* fibroblasts ('conditioned' medium), they took up acid hydrolases from this conditioned medium. In contrast, normal cells failed to internalise enzymes from medium conditioned by prior growth of Jane's cells. This uptake of acid hydrolases into Jane's cells had a pronounced effect on turnover of intracellular macromolecular constituents. This is illustrated in Figure 9.3 where her cells were grown in medium containing [^{35}S]sulphate to label mucopolysaccharides (MPS). The data in Figure 9.3 show what happened to the labelled MPS after labelling was discontinued.

Internalisation of normal enzymes by Jane's cells was examined in more detail. Table 9.2 shows the effects of including various sugars and their derivatives in the medium. In addition, internalisation was decreased by prior treatment of the conditioned medium with a phosphatase.

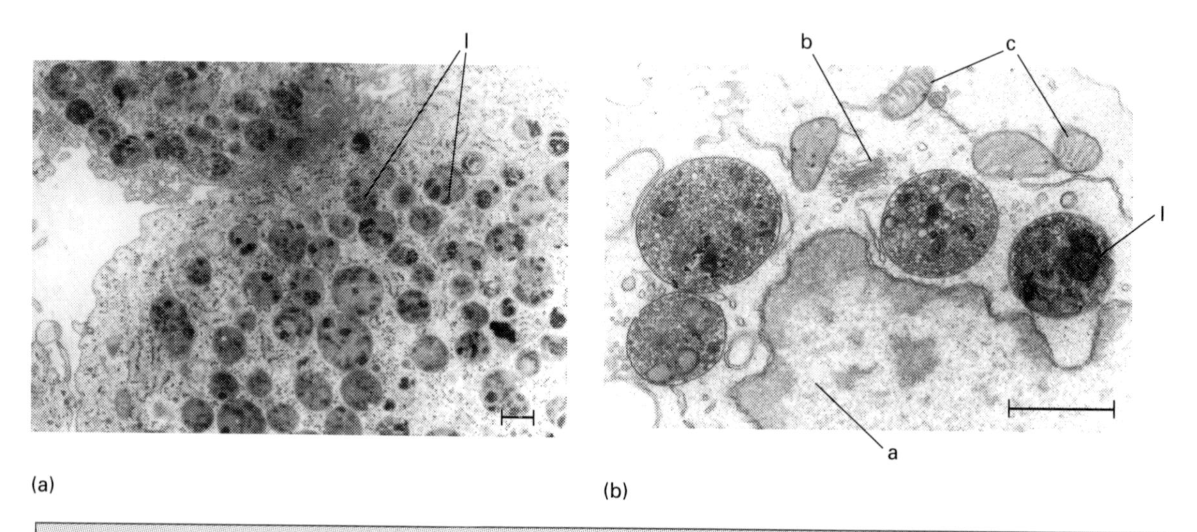

Figure 9.2. Electron micrographs of fibroblasts (panel a) and a monocyte (panel b) from the patient, showing inclusion bodies (I). Bars = 100 nm. Micrographs kindly supplied by Professor F. van Hoof (International Institute of Cellular and Molecular Pathology, Brussels).

Table 9.1. Distribution of β-glucuronidase between cells and medium.

Fibroblasts	In cells (U/mg protein)	In medium (U/ml)
Jane's	30	2000
Normal	178	11

Table 9.2. Uptake of β-glucuronidase (U/h) by cells.

Addition to medium	β-glucuronidase internalised
None	117
Glucose	120
Galactose	116
Fructose	92
Mannose	80
Fructose 6-phosphate	113
Glucose 6-phosphate	97
Galactose 6-phosphate	108
Mannose 6-phosphate	25
Methylmannoside	78
Polymannose	64

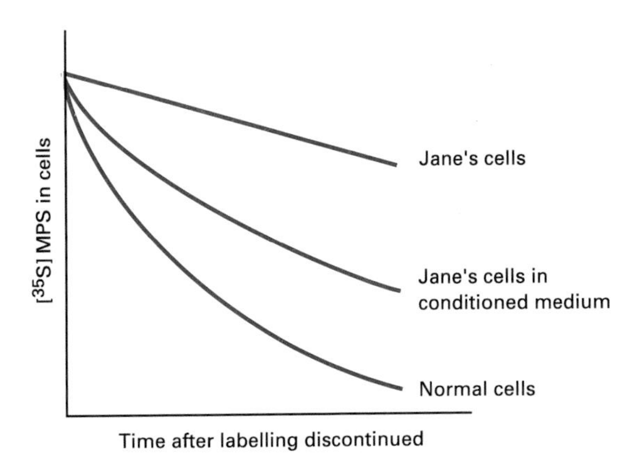

Figure 9.3. Turnover of labelled mucopolysaccharides.

Questions

1. What is the normal subcellular location of the acid hydrolases? Why are they referred to as acid hydrolases? What, collectively, is their role in normal cells?
2. What are the structures labelled a, b and c in Figure 9.2b?
3. What is the likely identity of the inclusion bodies in the cells? Why do they accumulate a variety of macromolecular components?
4. What can be concluded from the data concerning the distribution of the acid hydrolases between cells and medium?
5. Where in the cell does turnover of mucopolysaccharide take place? Explain the behaviour of Jane's fibroblasts before and after exposure to conditioned medium.
6. What is the likely mechanism involved in enzyme internalisation? Do the experimental data give any insights concerning its biochemical requirements?
7. What is the metabolic lesion in the patient?

Commentary

Jane suffers from the rare genetic disorder known either as *mucolipidosis type II* (from the accumulation of mucolipids along with macromolecular components) or as *I-cell disease* (from the inclusion bodies seen in the cells, Figure 9.2a). The biochemical lesion is the inability to produce nascent acid hydrolases with the correct intracellular address tag, mannose 6-phosphate, required for their intracellular routing to their destination, the lysosomes. Although a very rare condition, I-cell disease was crucial to our understanding of the ways in which cells sort newly-synthesised proteins.

The acid hydrolases are normally confined to the lysosomes. They are maximally active at the acidic pH (about 5) characteristic of the lysosomal interior. Collectively they catalyse the hydrolysis of a wide range of macromolecular constituents to small molecules during the normal turnover of cellular components. Cells of the patient secrete most of their acid hydrolases instead of sequestering them in the lysosomes. The inclusion bodies represent lysosomes that have become choked by accumulated macromolecules which cannot be degraded due to the lack of acid hydrolases. Figure 9.2b shows these enlarged lysosomes in the vicinity of the cell nucleus (a), the Golgi complex (b) and mitochondria (c).

Mucopolysaccharides (MPS), as representative examples of intracellular components requiring lysosomal degradation, are only poorly processed by the patient's cells. However, after these cells have internalised normal acid hydrolases from conditioned medium, MPS are degraded much more rapidly. This suggests two things. Firstly, Jane's cells can be 'corrected' by internalisation of normal lysosomal enzymes, and secondly that the loss of lysosomal proteins by her cells must be due to a defect in the enzymes rather than in the retention mechanism itself.

Internalisation occurs by binding of the hydrolases to surface receptors followed by their endocytosis via clathrin-coated pits. The endocytotic vesicles fuse with lysosomes to deliver their cargo of acid hydrolases. The surface receptors recognise a specific ligand attached to normal hydrolases whose identity can be deduced from competition studies (Table 9.2). Addition to the medium of an excess of a compound similar in structure to the natural address tag of the hydrolase will swamp the cell's surface receptors and prevent endocytosis. Table 9.2 shows that carbohydrates containing mannose, and especially mannose 6-phosphate, are the most effective competitors. This implies that the acid hydrolases are glycoproteins with oligosaccharides containing mannose 6-phosphate. Phosphatase treatment also inhibits internalisation, showing that the mannose residues must be in the phosphorylated state for the recognition system to work.

Nascent lysosomal proteins acquire their oligosaccharide components in the endoplasmic reticulum and Golgi complex. Their mannose 6-phosphate groups act as the lysosomal address tags. Normally most of these proteins are sent directly to the lysosomes and very little is lost by secretion. Why cells internalise lysosomal enzymes from the medium is not clear. Nevertheless, cell biologists were able to exploit this apparent curiosity to deduce the protein-sorting mechanism operating inside the cell.

It is now known that I-cell disease involves a genetic defect in the gene encoding the phosphotransferase required for the phosphorylation of the terminal mannose of the oligosaccharide. As expected, the effects of the deficiency are widespread and devastating. Death usually occurs through cardiorespiratory complications at 5–7 years of age. There is no treatment, but prenatal diagnosis is possible using amniotic cells.

There are now several inherited human diseases in which there is known to be a defect in protein targeting, including cystic fibrosis. The mutation most commonly responsible for cystic fibrosis is a single amino acid deletion in the transmembrane regulator protein which functions as a regulated channel for chloride ions. The mutant protein becomes trapped in the endoplasmic reticulum rather than reaching the plasma membrane and chloride transport is thereby disrupted. In familial hypercholesterolaemia, one particular mutation of the low density lipoprotein (LDL) receptor produces a defective receptor protein that can bind LDL but fails to be internalised. In hyperoxaluria, a mutation in the enzyme alanine-glyoxalate transaminase produces a wrong 'address' so that the protein gets delivered to mitochondria rather than peroxisomes!

Further Questions

1. Tay–Sachs disease (Appendix 4) is representative of the lysosomal storage disorders. How do these disorders differ from I-cell disease?
2. What are the 'advantages' to the cell of confining its acid hydrolases in the lysosomes? Give other examples and other advantages of cellular compartmentalisation.
3. What type of mechanism would be involved in maintaining the interior of the lysosome at an acid pH?
4. This problem concerned the correct sorting of proteins for delivery to an organelle but a few organellar proteins are synthesised *in situ*. Which are these and how is their synthesis brought about?
5. When cell biologists first turned their attention to the problem of understanding how proteins are selected for secretion, it was thought that the cell possessed two classes of ribosomes, one specialised for synthesis of cytosolic proteins and others for secretory proteins. What experiments could be done to show that ribosomes are indiscriminatory and that possession by a secretory protein of an *N*-terminal pre- or 'signal' sequence is the determining factor in its subcellular routing?

Connections

- Use this problem to review the different mechanisms by which substances enter and leave cells such as receptor-mediated endocytosis, exocytosis, transporters, ion channels, etc. What distinguishes active and passive mechanisms? (See also Problems 13 and 20.)
- Lysosomes are also involved in defence mechanisms. Review this topic and use it to find out something about how the bacteria responsible for leprosy and tuberculosis evade the body's surveillance.

- Ensure you understand what the term 'protein glycosylation' means. Where does glycosylation take place in the cell? List the various functional roles for protein glycosylation.
- Glycosylation is just one of many post-translational modifications which a protein can undergo. See how many others you can remember. (See also Problems 8 and 18.)
- Lysosomes contain proteinases which are involved in the cell's 'garbage disposal' system. However, the body employs proteinases in very many other areas of metabolism and control. To 'jog' your memory – zymogens/digestive enzymes, prohormones, collagen, blood clotting and complement, polyproteins – do these ring bells and can you add to the list? (See also Problems 8, 16 and 18.)
- This problem made diagnostic use of monocytes. Use this opportunity to recall the different types of blood cells and the diagnostic value of a blood test. (See also Problems 3, 5–7, 10–12, 18–20 and 22.)

References

Armstrong J (1992) Another protein out in the cold. *Nature* **358**, 709-10. *News and views* article on examples of mis-targeting in disease, including I-cell disease and cystic fibrosis.

Dice J F (1990) Peptide sequences that target cytosolic proteins for lysosomal proteolysis. *Trends in Biochemical Sciences* **15**, 305–9.

von Figura K (1991) Molecular recognition and targeting of lysosomal proteins. *Current Opinion in Cell Biology* **3**, 642–6.

Fukuda M (1991) Lysosomal membrane glycoproteins. Structure, biosynthesis, and intracellular trafficking. *Journal of Biological Chemistry* **266**, 21 327–30. Mini-review.

von Heijne G (1990) Protein targeting signals. *Current Opinion in Cell Biology* **2**, 604–8.

Paulson J C (1989) Glycoproteins: what are the sugar chains for? *Trends in Biochemical Sciences* **14**, 272–6.

10 ACID EXCRETORS

Introduction

Anaemia can be a symptom of most forms of malnutrition, but two vitamins are especially involved in the formation of blood components, vitamin B_{12} and folic acid. Vitamin B_{12} deficiency is nearly always due to a failure to absorb the vitamin. For it to be absorbed from the intestine, vitamin B_{12} has first to be bound to a special glycoprotein ('intrinsic factor') secreted by the stomach.

Pernicious anaemia, or Addisonian anaemia, was first described in 1847 by Thomas Addison of Guy's Hospital, London, as a general anaemia that was invariably fatal. It was later called 'pernicious anaemia', but remained intractable until 1926 when it was shown that feeding raw liver cured the disease. The curative factor in liver, the 'extrinsic factor', turned out to be vitamin B_{12}. In 1948 the vitamin was isolated and in 1955 its chemical structure was established by Dorothy Hodgkin in the first application of crystallography to a biologically important molecule. It is unusual in containing cobalt, hence its other name, cobalamin. Injection of pure cobalamin

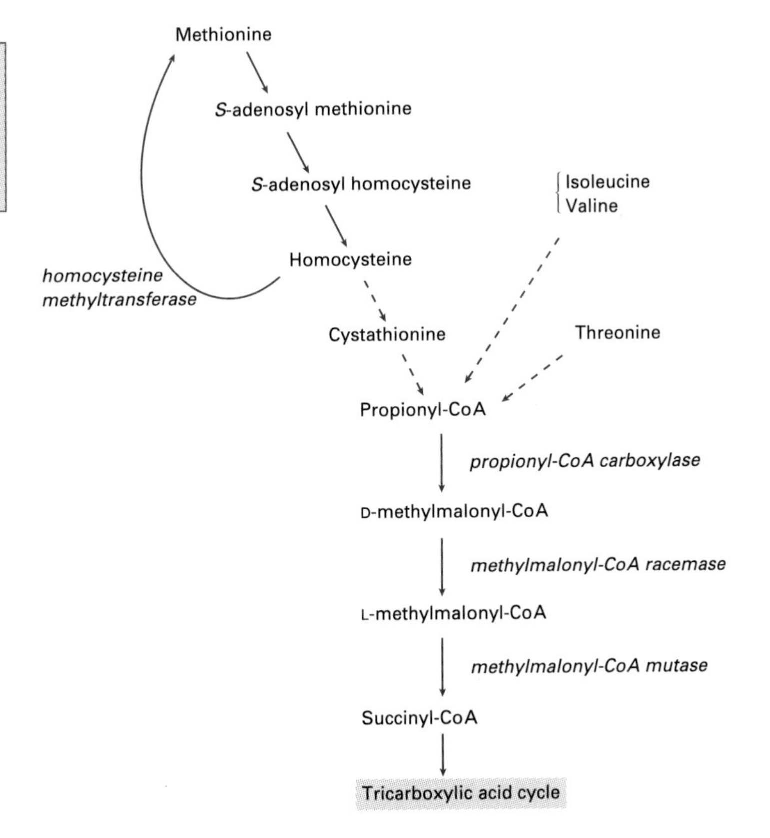

Figure 10.1. Degradation of amino acids to succinyl-CoA via propionyl-CoA and methylmalonyl-CoA.

completely controls pernicious anaemia. It is not known why the stomach stops producing intrinsic factor but an autoimmune reaction has been suggested.

Only two human enzymes require cofactors derived from vitamin B_{12} (Figure 10.1). Each requires a different form of cobalamin:

1. *Methylmalonyl-CoA mutase.* This catalyses an intramolecular rearrangement of methylmalonyl-CoA to succinyl-CoA. This represents the entry point for the carbon skeletons of several amino acids into the tricarboxylic acid cycle. Adenosylcobalamin is the form of vitamin B_{12} involved.
2. *Homocysteine methyltransferase.* This requires methylcobalamin and catalyses the methylation of homocysteine to methionine. The cobalamin allows the transfer of the methyl group from N^5-methyl tetrahydrofolate to homocysteine.

Not all patients with defects in vitamin B_{12} metabolism suffer from pernicious anaemia. This problem features another group of disorders involving this most unusual vitamin.

The Problem

A genetic condition known as methylmalonic acidaemia has been described. Patients with this condition have high concentrations of methylmalonic acid (in the form of methylmalonate) in their blood and as a result excrete large amounts of methylmalonate in their urine. However, in other respects they are a diverse population. Group A patients show ketoacidosis and their methylmalonate excretion is markedly decreased by intramuscular injection of pharmacological (very large) doses of vitamin B_{12} (Figure 10.2). Group B patients also have ketoacidosis but they fail to repond to vitamin B_{12}. Group C patients are distinguished by having homocystinuria, cystathioninuria and hypomethioninaemia. They also respond to vitamin B_{12} injections.

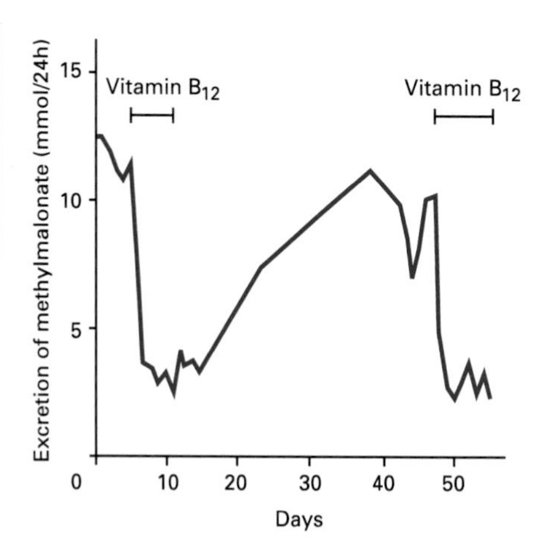

Figure 10.2. Injection of vitamin B_{12} decreases the methylmalonate excretion by Group A patients.

Questions

1. What are the principal effects of ketoacidosis?
2. Ketoacidosis may also be a symptom of diabetes mellitus. How would you ascertain that patients in Groups A and B were not diabetic?
3. In the metabolism of the branched-chain amino acids, valine and isoleucine, to methylmalonate, what is the first step likely to be?
4. Deduce where the metabolic blocks are in the patients of Groups A and B. Why do Group A but not Group B patients respond to vitamin B_{12}?
5. What is the metabolic origin of the homocystine in the urine of Group C patients?
6. Where is the metabolic block in Group C patients?
7. What other areas of metabolism might also be affected in Group C patients?

Commentary

The first step in the degradation of valine and isoleucine is their transamination catalysed by the branched-chain amino acid transaminase (aminotransferase). Their carbon skeletons are then degraded to propionyl-CoA (Figure 10.1).

Ketoacidosis occurs in a number of situations including starvation and uncontrolled diabetes mellitus. In the cases described in this problem, large amounts of an organic acid, methylmalonic acid, enter the blood. Depending on its extent, metabolic acidosis can lead to life-threatening electrolyte imbalance and dehydration.

The principal buffer of blood is the bicarbonate/carbonic acid system:

$$H^+ + HCO_3^- \leftrightarrow H_2CO_3$$

Addition of the ketoacid (symbolised as H^+A^-) to the plasma shifts the equilibrium towards H_2CO_3, thus decreasing $[HCO_3^-]$ and increasing $[H_2CO_3]$:

$$H^+A^- + HCO_3^- \rightarrow H_2CO_3 + A^-$$

The blood pH will fall since by the Henderson–Hasselbalch equation:

$$\mathrm{pH} = \mathrm{p}K + \log\frac{[HCO_3^-]}{[H_2CO_3]}$$

This fall in pH stimulates respiration and the increased ventilation 'blows off' CO_2 thereby decreasing $[H_2CO_3]$:

$$H_2CO_3 \rightarrow CO_2 + H_2O$$

so the $[HCO_3^-]/[H_2CO_3]$ returns towards the original value and with it the pH. However, there is a limit to which respiration can be increased.

In addition to this respiratory compensation mechanism, we must also consider the action of the kidneys. These excrete the non-volatile ketoacid anion and H^+, but are limited in the extent to which they can excrete H^+ (the lower limit of urinary pH is about 4.8). At the same time the excreted ketoacid anion A^- must be accompanied by a cation. Initially Na^+ and K^+ serve this function but this cannot continue for long or there would be a serious loss of electrolytes. To overcome these problems the kidney forms NH_3 which combines with H^+ to form NH_4^+ allowing the ketoacid anion to be excreted as its ammonium salt. This mechanism does two things. It substitutes NH_4^+ as the cation, thereby sparing Na^+ and K^+, and at the same time acts as a buffer allowing the excretion of large amounts of H^+. Even so there is a limit and in severe metabolic acidosis Na^+ and K^+ have to be sacrificed. Since renal reabsorption of H_2O is linked to the transtubular movement of Na^+, there is also a net loss of H_2O and consequent dehydration.

Figure 10.3. Glucose tolerance test.

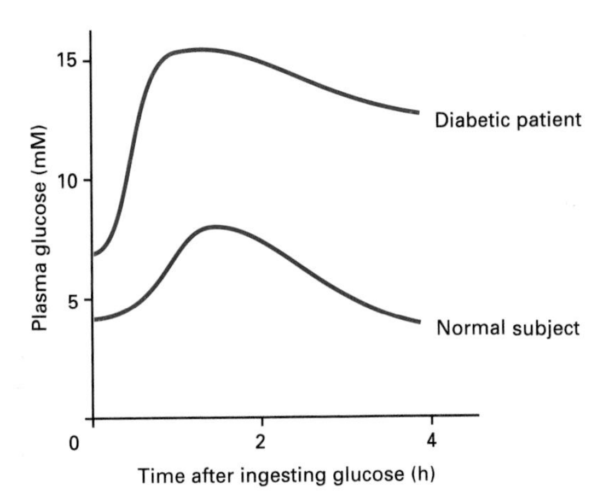

Diabetes is characterised by an inability to handle glucose efficiently as a result of insulin insufficiency or resistance. There is fasting hyperglycaemia. An intake of glucose in a meal or in the glucose tolerance test (Figure 10.3), which is a provocative procedure designed to measure glucose handling, would result in a large and sustained rise in plasma glucose. Since this usually exceeds the renal reabsorption threshold for glucose (about 7–10 mM), glucosuria ensues. Water follows the excreted glucose resulting in polyuria and hence polydipsia.

A metabolic block at either the methylmalonyl-CoA mutase or the racemase would lead to both methylmalonic acidaemia and methylmalonic aciduria. Genetic conditions such as this, which show Mendelian inheritance, result from a mutation in a single gene. We can be more precise about the location of the defect in Group A patients because they respond to vitamin B_{12}. Since the mutase, but not the racemase, requires adenosylcobalamin as a cofactor, we may speculate that this group has a block at the mutase due to a problem with the adenosylcobalamin cofactor, possibly its synthesis from dietary vitamin B_{12}. In the case of Group B patients, the evidence presented suggests that the problem is not with the adenosylcobalamin cofactor. While this might cause us to suggest a block at the racemase, remember that the mutation might affect the mutase apoprotein. In fact in the Group B methylmalonic acidaemia cases so far investigated, mutations in the mutase gene predominate.

Group C patients have low plasma methionine levels and excrete both homocystine and cystathionine. Homocystine consists of two molecules of homocysteine linked by a disulphide bond. These observations point to a deficiency in the enzyme homocysteine methyltransferase, but remember there must also be a block at the mutase or racemase to give rise to the methylmalonic acidaemia. Since the mutase and the methyltransferase both require cofactors derived from vitamin B_{12}, albeit different forms, and Group C patients respond to treatment with vitamin B_{12}, the most likely explanation for this type of methylmalonic acidaemia is a general defect in vitamin B_{12} metabolism which affects both B_{12}-dependent enzymes.

As mentioned in the Introduction to this problem, pernicious anaemia results from a general deficiency in vitamin B_{12} due to malabsorption. So are Group C patients with methylmalonic acidaemia really suffering from pernicious anaemia? Certainly pernicious anaemia patients would be expected to excrete methylmalonic acid and some of the Group C patients (but none of Groups A and B) have also been shown to have megaloblastic anaemia, a hallmark of pernicious anaemia. However, it must be stressed that the two disorders, pernicious anaemia and methylmalonic acidaemia, are quite distinct conditions.

Homocysteine methyltransferase serves to remethylate methionine and participates in the active methyl cycle involving *S*-adenosyl methionine. Consequently we would expect many other methylation reactions dependent on this active methyl cycle to be adversely affected in Group C patients.

Methylmalonic acidaemia has an incidence of about 1 in 29 000, of which about half are in Group B, but the true incidence may be significantly higher since some neonatal deaths from this disorder may go unrecognised (see the article by Hoffman in the References). Conventional treatment involves dietary restriction and this can provide a relatively normal life. Nonetheless there is usually a degree of mental retardation and even well-managed patients are at risk of life-threatening episodes of metabolic acidosis associated with infections or dietary lapses. Gene replacement therapy for Group B patients is a long-term goal.

Further Questions

1. Some patients with methylmalonic acidaemia may be treated with vitamin B_{12}. Where this is inappropriate, methylmalonic acid production has been controlled by dietary restriction of protein. What is the rationale of this treatment and what are some of the problems that might be encountered?
2. Transaminases (aminotransferases), such as the one which catalyses the first step in the metabolism of isoleucine and valine, and carboxylases, including propionyl-CoA carboxylase (Figure 10.1) also require cofactors. In each case, what is the cofactor?
3. How is the NH_3 formed in the kidney, and what is its source?
4. A genetic condition has been described called multiple carboxylase deficiency in which there is a generalised deficiency in the cofactor required by carboxylases. What areas of metabolism would be affected?
5. From the information given in this problem, what treatment would you include in the long-term management of a patient who has had a partial or complete gastrectomy?
6. Normally a single gene defect would not be expected to affect more than one enzyme and so only one step of a metabolic pathway would be blocked. However, exceptions to this are known. An example is orotic aciduria where two enzymes of the pyrimidine biosynthetic pathway, orotidylate pyrophosphorylase and orotidylate decarboxylase, are both deficient. In what ways could such a situation arise, apart from effects on a common cofactor as in the Group C methylmalonic acidaemia patients described here?

7. If the blood concentrations of HCO_3^- and H_2CO_3 are 25 mM and 1.2 mM, respectively, and the pK of the bicarbonate/carbonic acid dissociation is 6.1, what is the pH of the blood?
8. If the minimum attainable urinary pH is 4.8, what volume of urine would be required to excrete a daily production of 50 mmol of acid as free H^+?

Connections

- Use this problem to review your basic knowledge of pH and buffers, including acid–base balance in the body and the role of the kidney. (See also Problems 1, 5, 7 and 19.)
- Review what you know about vitamins. Define the term 'vitamin' and list the vitamins, divided into water-soluble and lipid-soluble groups. Is vitamin D a vitamin or a hormone?
- Methionine is a special amino acid in a number of ways. What are the essential structural and functional features of this amino acid? It plays an important role in protein biosynthesis when mRNA is translated by the ribosome and it is involved in the transfer of methyl groups. Ensure you are familiar with both these areas.
- In the metabolic degradation of amino acids, some are said to be glucogenic while others are ketogenic. Name examples of each and explain the difference.
- List the different ways in which inborn errors of metabolism, such as methylmalonic acidaemia, may be treated or the symptoms alleviated. What would be needed to cure such diseases?
- Check that you understand the various metabolic states in which ketoacidosis can arise and what the biochemical basis is.
- Besides the cobalt in vitamin B_{12}, metals are important components of other biological compounds. Name the metals in myoglobin, cytochrome oxidase, chlorophyll, xanthine oxidase, carboxypeptidase A, lysyl oxidase and pyruvate carboxylase. Review the roles of these.

References

Green A (1989) In-born errors of organic acid metabolism. *British Journal of Hospital Medicine* **41**, 426–34.

Hoffman M (1991) Scientific sleuths solve a murder mystery. *Science* **254**, 931.

Ledley F D (1990) Perspectives on methylmalonic acidaemia resulting from molecular cloning of methylmalonyl CoA mutase. *Bioessays* **12**, 335–40.

Rosenblatt D S and Cooper B A (1990) Inherited disorders of vitamin B_{12} utilization. *Bioessays* **12**, 331–4.

11 FAILURE TO STOP

Introduction

Not only do we have an enormous accumulated knowledge of haemoglobin as a *protein*, but we also know a very great deal about the *genes* that code for haemoglobin. This means in turn that a very large number of mutations in the haemoglobin genes have been identified. The majority of these mutations are harmless but some cause severe or even fatal consequences (the haemoglobinopathies). The best characterised haemoglobinopathy is sickle cell anaemia, the first human single gene defect fully understood at the molecular level. In fact, sickle cell disease is rather unusual. The defect leads to the incorporation of a hydrophobic amino acid, valine, in the β-globin chain in place of a polar glutamate residue. This makes the haemoglobin molecule relatively more 'sticky' and this aggregation of haemoglobin in erythrocytes leads to sickling and haemolysis.

Collectively the haemoglobinopathies constitute the largest group of single gene disorders in the human population. One large subgroup of these is the thalassaemias, so-called because they were originally identified in Mediterranean peoples ('thalassa' means 'sea', and to the Greeks and Romans, the Mediterranean was *the* sea). In fact, thalassaemias are very widely distributed across Africa, the Mediterranean region, the Indian subcontinent and South-East Asia (Figure 11.1). The disease is also found in

Figure 11.1. World distribution of the thalassaemias. Redrawn from Weatherall and Clegg (1981).

migrant populations from these regions and sickle cell anaemia in the black population of the USA is a legacy of the Slave Trade.

In many haemoglobinopathies single amino acid mutations in the globin chains result in a failure to bind haem or carry oxygen properly. In contrast, thalassaemias are characterised by an imbalance in the synthesis of the α- and β-globin chains. The occurrence of these mutations in areas where malaria is endemic has led to the suggestion that heterozygous individuals have a greater resistance to malaria, although homozygous individuals pay the penalty in increased morbidity and mortality.

The Problem

Suzy W., a 15-year-old girl of Chinese origin, had a history of lassitude. Her sclerae were yellow and her spleen was palpable. A haematological examination showed that her erythrocytes were rather hypochromic (pale) with anisocytosis (variation in size) and poikilocytosis (variation in shape) (Figure 11.2). Staining revealed reticulocytes and some of the erythrocytes contained inclusion bodies (Figure 11.3). Table 11.1 shows the quantitative blood analysis.

The haemoglobins (Hb) present in Suzy's blood were separated by electrophoresis at pH 8.6 on cellulose acetate and a number of abnormal haemoglobins were found (Figure 11.4).

Suzy's haemoglobins were assayed for α- and β-globins. Hb 1 (Figure 11.4) possessed both α- and β-globin chains but Hb 2 contained only β-globin. Overall, the ratio of α- and β-globins in the patient was 0.59:1.0 (normal = 1.03:1.0).

Samples of α-globin from Hb 1 and from Hb A, the major haemoglobin of normal blood, were digested with trypsin (an endoprotease which cleaves polypeptide chains on the *C*-terminal side of basic amino acid residues) and their tryptic peptides were separated by two-dimensional electrophoresis and paper chromatography (Figure 11.5).

The α-globin from Hb 1 was also treated with cyanogen bromide (CNBr), which cleaves proteins on the *C*-terminal side of methionine residues. A number of CNBr peptides were produced, the largest of which was approximately 50% larger than its equivalent in Hb A.

Table 11.1. Quantitative blood analysis.

	Suzy W.	Normal range (female)
Haemoglobin (g/dl)	8.6	11–15
Mean corpuscular haemoglobin (pg/cell)	22	26–33
Red cell count (per litre)	4.4×10^{12}	$3.9–5.6 \times 10^{12}$
Reticulocytes (%)	5.5	0.5–2.5

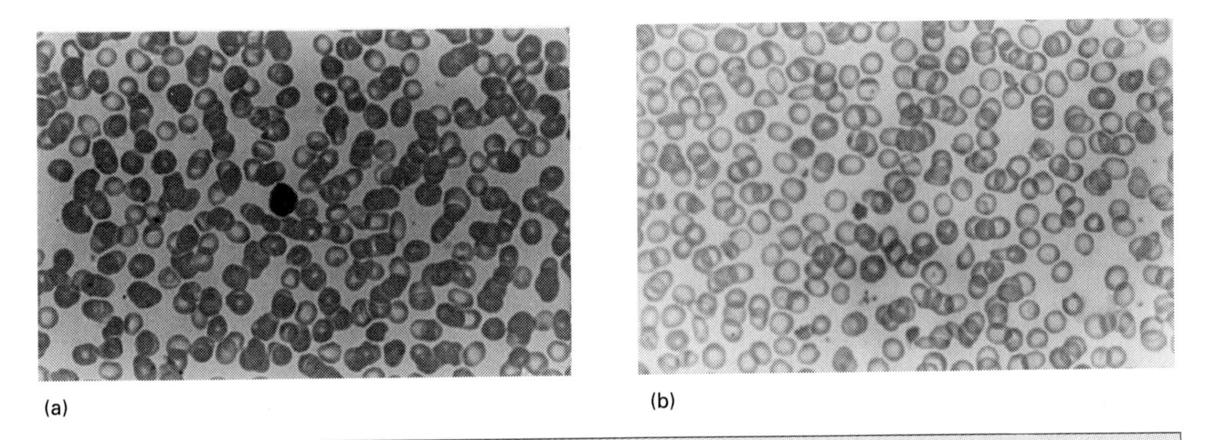

Figure 11.2. Peripheral blood films (normal subject, panel (a); patient, panel (b)). Photographs kindly supplied by Ann Urmstom, Manchester Royal Infirmary.

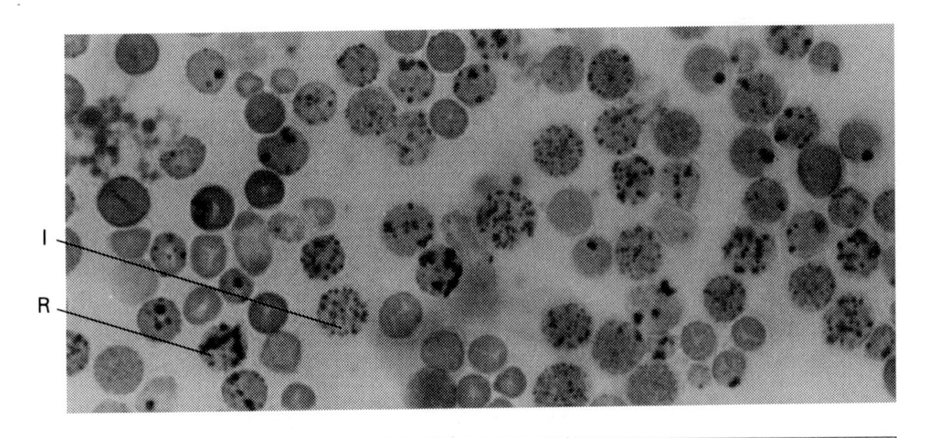

Figure 11.3. Blood film of the patient, stained to show reticulocytes (R) and cells with inclusion bodies (I). Photograph kindly supplied by Ann Urmstom, Manchester Royal Infirmary.

Figure 11.4. Electrophoresis of haemoglobins in cellulose acetate at pH 8.6.

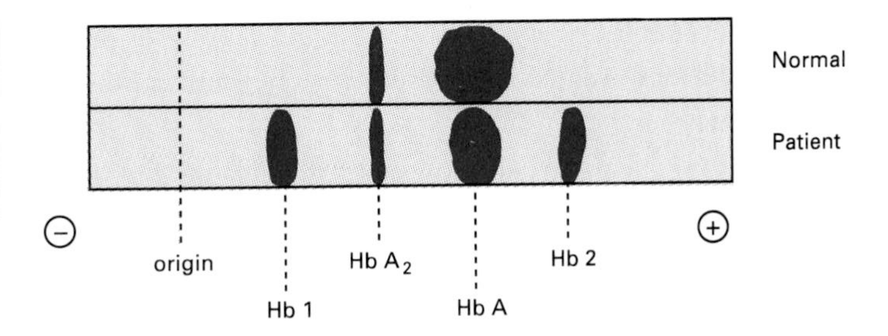

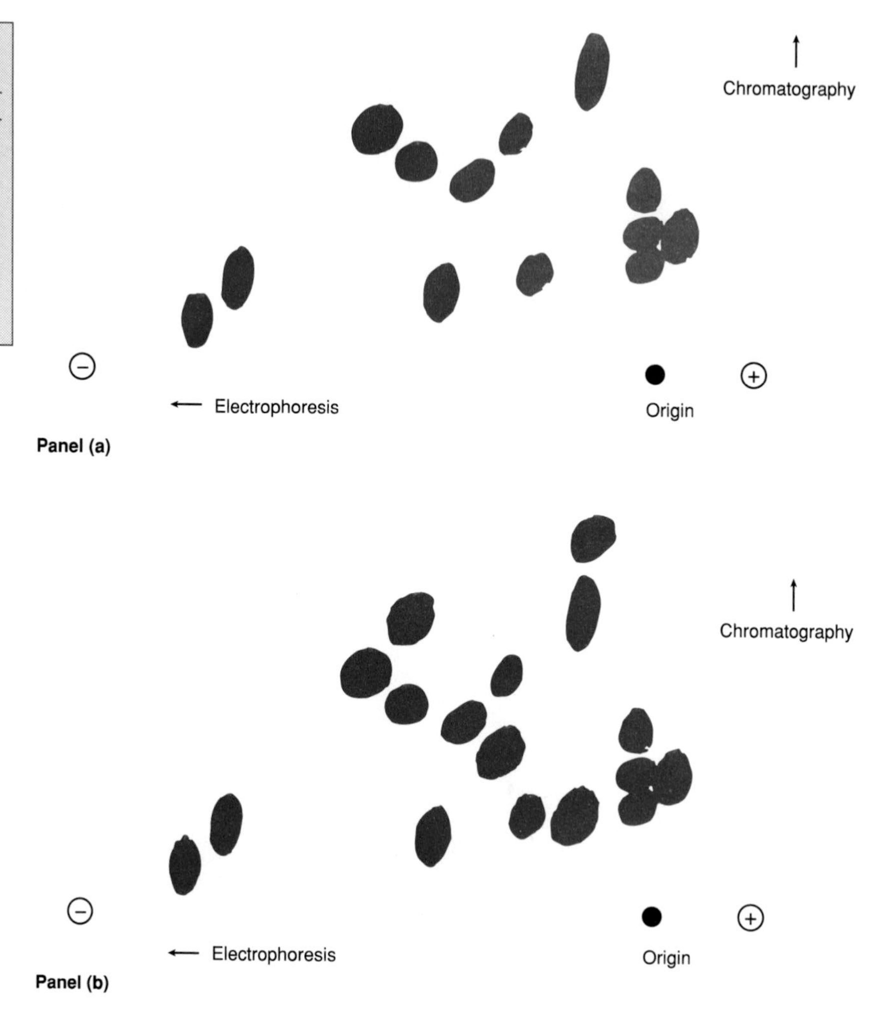

Figure 11.5. Two-dimensional separation of tryptic peptides of the α-chains of haemoglobins. Panels (a) and (b) show the peptides from Hb A and Hb 1, respectively. Based on data in Clegg *et al.* (1971) *Nature* 234, 337–40.

Questions

1. What do the presenting symptoms, the initial examination and the results of the blood analysis (Figures 11.2 and 11.3, Table 11.1) suggest to you about the patient's clinical condition?

2. What can you conclude from the examination of the patient's haemoglobins (Figure 11.4)?

3. What do the trypsin digestion data (Figure 11.5) indicate about the relationship between the α-globin chains of Hb A and Hb 1?

4. Refer to Figure 11.6, which shows the amino acid sequence of human α-globin together with the nucleotide sequence of its mRNA. Where is the largest CNBr

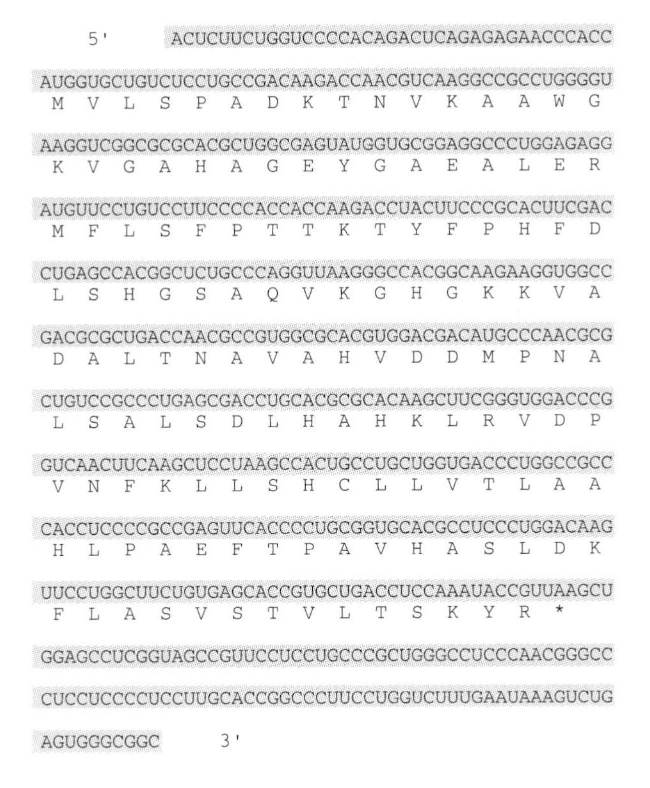

Figure 11.6. The nucleotide and amino acid sequences for human α-globin. The single-letter abbreviations for the amino acids will be found in Appendix 5 and the Genetic Code is given in Appendix 6. Termination of the polypeptide chain is indicated by *.

fragment of α-globin located in the amino acid sequence? You have been told that this CNBr fragment is 50% larger in Hb 1 than in Hb A. Bearing in mind the mRNA sequence, how might this be explained? What then is the amino acid sequence of the α-globin in Hb 1 likely to be?

5. How does your suggestion for the amino acid sequence of the α-globin in Hb 1 explain the differences in the tryptic peptide maps for Hb A and Hb 1 (Figure 11.5)?

6. What parameters determine the electrophoretic mobility of a protein in cellulose acetate (Figure 11.4)? Does the sequence which you have suggested for the α-globin of Hb 1 explain the relative electrophoretic mobilities of Hb A and Hb 1?

7. What is the structure of Hb 2 (Figure 11.4) and how would it have been formed?

8. How does your suggestion for the molecular defect leading to Hb 1 and Hb 2 explain Suzy's symptoms and her haematology?

Commentary

The condition described here is a type of thalassaemia in which a mutation causes an imbalance in the relative amounts of the different globin chains. The most common thalassaemias are due to a deficiency of either α-globin (α-thalassaemia) or β-globin (β-thalassaemia). The affected globin may be essentially absent (α^0- and β^0-thalassaemias), usually due to a gene deletion, or its synthesis may be at a reduced rate (α^+- and β^+-thalassaemias) as a result of a variety of mutations. These may affect gene transcription, RNA processing, mRNA translation or globin stability.

Suzy's clinical picture is of anaemia; her total haemoglobin is low, there is less haemoglobin per cell, and the erythrocytes are hypochromic. This may be due to haemolysis since there is jaundice (yellow sclerae) and splenomegaly. Erythrocytes normally have a lifespan of about 120 days before they are removed from the circulation and destroyed in the spleen. Their haemoglobin is turned over. Whereas the iron is conserved, the globin and haem are degraded. Increased haem degradation results in elevated plasma levels of bilirubin which may be deposited in the peripheral tissues (hence the jaundiced sclerae). Why is there increased haemolysis? The spleen acts as a 'filter' removing abnormally-shaped erythrocytes from the circulation. Suzy's blood contains a higher than normal proportion of such cells, including cells with inclusion bodies (Figure 11.3). The relative hypoxia of anaemia leads to compensatory erythropoiesis so that a larger than usual proportion of immature red cells, the reticulocytes (Figure 11.3, Table 11.1), is released into the peripheral circulation.

Normal blood contains mainly Hb A ($\alpha_2\beta_2$) with some Hb A_2 ($\alpha_2\delta_2$). Electrophoresis (Figure 11.4) revealed that Suzy's blood contains these normal haemoglobins but also two abnormal haemoglobins, here designated Hb 1 and Hb 2. Although Hb 1 was shown to contain both α- and β-globins, the tryptic map of its α-globin (referred to here as $\alpha^\star$-globin) is abnormal. It contains four peptides not found in α-globin, whereas all the other tryptic peptides are common to both α- and $\alpha^\star$-globins (Figure 11.5). A tryptic map is sometimes called a 'fingerprint' since it is unique to the protein from which it is derived. Thus, we can conclude two things from Figure 11.5: that $\alpha^\star$-globin contains the *same* amino acid sequence as α-globin but in addition possesses an *extra* amino acid sequence to give rise to its four unique tryptic peptides. The CNBr cleavage provides the clue as to where in the polypeptide chain this additional sequence lies. The largest CNBr fragment of α-globin comprises the *C*-terminal 65 amino acids (see Figure 11.6). Since this is 50% larger in $\alpha^\star$-globin, $\alpha^\star$-globin must comprise α-globin with a *C*-terminal extension of about 30–35 residues. This extension has to include an additional three basic residues to provide, along with the former *C*-terminal arginine, the four extra trypsin cleavage sites to account for the four $\alpha^\star$-globin-specific tryptic peptides.

How can we explain the *C*-terminal extension of $\alpha^\star$-globin? The simplest explanation is a point mutation in the stop codon (UAA) of the α-globin mRNA, changing it to a sense codon. This would allow the ribosomes to run on into what is normally the 3′ untranslated region of α-globin mRNA and continue translating the mRNA in frame. After an additional 31 amino acids, another stop codon (also UAA) is reached (Figure 11.7). This extra *C*-terminal extension includes three arginine residues which are sites for trypsin cleavage. An alternative type of mutation, namely the complete deletion of the normal stop codon, would also result in a *C*-terminal extension with the required characteristics. However, disruption of the stop codon by the insertion or deletion of a base can be dismissed since both of these would change

```
GUGAGCACCGUGCUGACCUCCAAAUACCGUCAAGCUGGAGCCUCGGUA
 V  S  T  V  L  T  S  K  Y  R  Q  A  G  A  S  V

GCCGUUCCUCCUGCCCGCUGGGCCUCCCAACGGGCCCUCCUCCCCUCC
 A  V  P  P  A  R  W  A  S  Q  R  A  L  L  P  S

UUGCACCGGCCCUUCCUGGUCUUUGAAUAAAGUCUGAGUGGGCGGC  3'
 L  H  R  P  F  L  V  F  E  *
```

Figure 11.7. Mutation in α-globin mRNA leading to the *C*-terminal extension of α-globin. The U → C point mutation in the normal stop codon UAA and the amino acids of the *C*-terminal extension are in bold type. (See Appendices 5 and 6.)

the reading frame downstream and create a *C*-terminal extension that either is too short or contains too few trypsin sites to explain the extra tryptic peptides. Hb 1 is known as haemoglobin Constant Spring (Hb CS), named after a Jamaican village – the home of the first affected family described. The mutation in the α-globin gene is now known to change the stop codon UAA to a glutamine codon CAA.

The $\alpha^\star$-globin chains combine with β-globin to form Hb 1 (structure $\alpha_2^*\beta_2$) which is both larger and more basic than Hb A, thus decreasing its anodal mobility in electrophoresis (Figure 11.4).

It might be expected that Hb CS, like many abnormal proteins, would be degraded more rapidly than Hb A. However, David Weatherall and John Clegg, the discoverers of the Hb CS mutation, found that this was not the case. What may happen is that the Hb CS α-globin mRNA is unstable and more readily degraded. There is some evidence that other mutations in the non-coding part of α-globin mRNA result in very little mRNA appearing in the cytoplasm.

The excess β-globin chains combine to form the other abnormal haemoglobin seen in Figure 11.4, Hb 2. These β_4 tetramers are actually known as haemoglobin H. Having a higher affinity for O_2 than Hb A, Hb H is relatively poor in delivering O_2 to the tissues, thus contributing to tissue hypoxia. It also tends to precipitate in the erythrocytes, especially as they age, resulting in the inclusion bodies (Figure 11.3), alterations in cell shape (Figure 11.2) and premature removal of the erythrocytes by the spleen.

In this patient the thalassaemia is relatively mild and requires no major treatment. Splenectomy would be indicated for severe haemolysis to prevent excessive premature destruction of erythrocytes. More severely affected thalassaemics may require blood transfusions. There is no permanent cure, but gene replacement therapy is a reasonable long-term goal.

Further Questions

1. Why are α-thalassaemic patients usually less severely affected than those with β-thalassaemia? (To answer this question you will need to find out more about the human globin genes.)

2. What would the haemoglobin electrophoresis (Figure 11.3) look like in β^{+}-thalassaemia?

3. What problems would be encountered with long-term treatment of thalassaemic patients by blood transfusion? How could they be ameliorated?

4. What types of haemoglobin are present in a normal foetus? What would be seen on electrophoresis of blood from a foetus carrying the Hb CS mutation?

5. How would you screen a foetus for the Hb CS mutation?

6. What are the differences in the symptoms of sickle cell disease and the type of thalassaemia described in this problem? Can any effective treatment be offered to sickle cell patients?

Connections

- Use this problem to review your understanding of the functions of haemoglobin. Check that you understand in general terms the relationship between amino acid sequence, protein structure and protein function.
- Find out about other haemoglobinopathies and learn how they have contributed to our understanding of haemoglobin function. What are their medical consequences and how can they be treated?
- In this problem trypsin was used as an analytical tool to create a peptide 'fingerprint'. List other proteinases that you know. Find out about their roles in digestion, intracellular degradation, prohormone activation, blood clotting and the complement cascade. (See also Problems 8, 9, 16 and 18.)
- Summarise what you know about gene structure. Do you understand what 'gene expression' means? How is the 'language' of a nucleotide sequence translated into 'amino acid language'? We all make many mistakes in our lives but the cell does not normally make errors in gene expression; recall how this is ensured?
- This problem featured the mutation of a single nucleotide and resulted in an extended polypeptide chain. Make a list of other types of mutation that can occur and their consequences. Check up on DNA repair mechanisms. (See also Problems 8, 18–20 and 22.)
- This problem used electrophoresis in cellulose acetate to separate haemoglobins. This form of electrophoresis is relatively easy to perform and the haemoglobins were easily recognised by their colour. Consequently, the other proteins that are present in the sample did not interfere with the analysis. What methods could have been used to reveal the presence of these other proteins? How would you detect a specific protein, such as an enzyme, after electrophoresis? Electrophoresis in cellulose acetate is a form of 'native' or non-denaturing electrophoresis as distinct from 'denaturing' electrophoresis in the presence of sodium dodecylsulphate (SDS). Make sure you understand what denaturation involves, why SDS is a denaturant and under what circumstances you would use 'native' and 'denaturing' conditions. (See also Problems 3, 8, 18, 20 and 22.)

- Haemoglobin is contained within the erythrocytes, but blood plasma also contains many proteins. Recall what these are and their functions. (See Problems 6, 12, 16–18 and 20.)
- Check that you understand how genetic disorders are diagnosed before birth and that you understand the terms 'hybridisation', 'probe', 'restriction enzyme' and 'blot'. (See Problems 8, 19 and 20.)
- Make sure you understand how haem is degraded in the body and the relationship between bilirubin and jaundice. Explain what happens in bruising. (See Problems 6 and 17.)
- Remind yourself of the differences in the oxygen-binding properties of haemoglobin and myoglobin, making sure you understand how these are related to the different functions performed by each protein. Ensure that you understand why the major haemoglobin in a foetus is different from Hb A. The switch-over from foetal to adult haemoglobin is completed shortly after birth. Explain why this can lead to neonatal jaundice, especially if the birth is premature, and what phototherapy does in these cases. (See also Problem 5.)

References

Davies K E and Read A P (1992) *Molecular Basis of Inherited Disease.* IRL Press/Oxford University Press, Oxford.

Verma I M (1990) Gene therapy. *Scientific American* **263**(5), 34–40.

Watson J D, Gilman M, Witkowski J and Zoller M (1992) DNA-based diagnosis of genetic diseases, Chapter 27, pp 539–66 *and* Working toward human gene therapy, Chapter 28, pp 567–81. In *Recombinant DNA.* Scientific American Books, New York.

Weatherall D J and Clegg J B (1981) *The Thalassaemia Syndromes* (3rd edn). Blackwell Scientific, Oxford.

Weatherall D J (1991) *The New Genetics and Clinical Practice.* Oxford University Press, Oxford, New York and Tokyo. Several useful chapters.

Weissman S M (1992) Gene therapy. *Proceedings of the National Academy of Sciences (USA)* **89**, 11 111–12.

12 GETTING THE MEASURE OF HORMONES

Introduction

Many hormones, growth factors and other biologically active agents are extremely potent, eliciting their effects at exceedingly low concentrations. For instance, the hormone glucagon (Fig. 12.1), the subject of this problem, is normally present in plasma at about 3×10^{-11} M or 100 pg/ml (1 pg $= 10^{-12}$ g). The concentrations in plasma or serum of some hormones are given in Appendix 4.

Clinical investigation frequently requires the rapid and accurate measurement of the concentrations of these hormones. Bioassays, which measure the hormone's effects in animals or organ systems, are expensive, slow and insensitive. Often they fail to distinguish between hormones whose biological effects are rather similar; for instance, there are several hormones that will increase blood glucose or cause bone growth. Antibodies provide the basis for a large variety of modern assay kits for compounds of clinical importance. They are quick and easy to use, extremely accurate, sensitive and specific.

Confirmation of human pregnancy is based on the detection of human chorionic gonadotropin (hCG) in the urine of a pregnant woman. This used to involve a bioassay – the gonadotropic effects of hCG were detected in toads – conducted by specialist laboratories. Not only was this expensive, but its insensitivity, lack of reproducibility and the time taken to obtain the results meant that pregnancy could only be confirmed once it was fairly well established. By using antibodies to hCG, highly accurate and rapid pregnancy testing can be done now in the privacy of the home using simple and inexpensive 'over-the-counter' kits.

Figure 12.1. Amino acid sequence of human glucagon.

His – Ser – Gln – Gly – Thr – Phe – Thr – Ser –Asp –Tyr –

Ser – Lys – Tyr – Leu – Asp – Ser – Arg – Arg – Ala – Gln –

Asp – Phe – Val – Gln – Trp – Leu – Met – Asn – Thr

The Problem

A clinical biochemistry laboratory established a radioimmunoassay (RIA) for human glucagon. A series of incubation tubes was set up, each tube containing a constant but limiting amount of an antiserum against human glucagon and a fixed amount of radiolabelled glucagon. Various amounts of unlabelled glucagon were also added. After incubation, the radioactive immune complex formed was assayed (Figure 12.2). The RIA was then used to assay plasma from three individuals (Table 12.1). Alan G. was a young child with a history of repeated convulsive limb movements. His parents reported that these seizures could be lessened by frequent feeding though this was becoming less effective. Barbara W. was a somewhat overweight post-menopausal woman who had been referred by a dermatology clinic because of her characteristic skin lesions. Fred C. was asymptomatic and had been screened as an apparently normal member of the general public. All three provided fasting blood samples for analysis (Table 12.2). Both Alan G. and Fred C. had a normal blood glucose response after injected glucagon (Barbara W. was not tested).

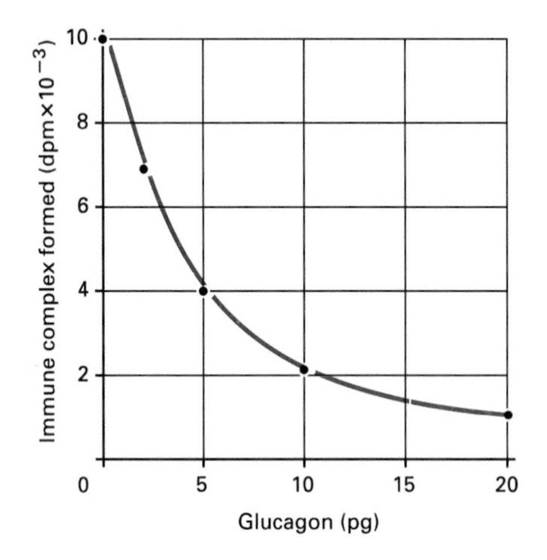

Figure 12.2. RIA standard curve for human glucagon. The assay measured the quantity of radioactive immune complex formed in the presence of the indicated amounts of purified human glucagon.

Table 12.1. Results of the RIA with the subjects' plasma samples.

Patient	Volume of plasma assayed (ml)	Immune complex formed (dpm)
Alan G.	0.10	3700
Barbara W.	0.01	2600
Fred C.	0.01	6400

Table 12.2. Analysis of fasting plasma glucose and insulin.

Patient	Glucose (mM)	Insulin (μU/ml)
Alan G.	2.4	9.6
Barbara W.	6.5	30.1
Fred C.	4.1	12.8
Normal subjects (range)	3.5–5.5	9.0–18.0

Questions

1. What are the plasma glucagon concentrations for each subject? The range for a group of normal subjects was found to be 85–150 pg/ml.
2. Why was it necessary to use a limiting amount of antiserum in the RIA?
3. Why was a *group* of individuals used for the control data?
4. What is the reason for Alan G.'s convulsive seizures and why were they relieved by feeding?
5. What is the metabolic defect in Alan G.?
6. What is the likely explanation for Barbara W.'s condition?
7. Fred C.'s plasma glucagon was tested again with the same result. Bearing in mind which property of glucagon is being measured in the RIA, how can his asymptomatic condition be reconciled with his plasma glucagon?
8. What treatments could be offered to Alan G. and to Barbara W.?

Commentary

The RIA is based on the ability of antibodies in the antiserum to recognise human glucagon with high specificity. These antibodies (Ab) form an immune complex with the radiolabelled glucagon (G*) according to the following equilibrium:

$$Ab + G^{\star} \leftrightarrow Ab\text{–}G^{\star}$$

The radiolabelled complex (Ab–G*) can be assayed by separating it from the remaining uncomplexed radiolabelled glucagon. Provided the Ab is present in limiting amounts, the presence of additional *unlabelled* glucagon (G) will result in competition between G and G* for Ab:

$$G + G^{\star} + Ab \leftrightarrow Ab\text{–}G^{\star} + Ab\text{–}G$$

The amount of radiolabelled complex will therefore decrease in proportion to the amount of unlabelled glucagon added as seen in Figure 12.2. Using this standard curve the plasma glucagon concentrations can be calculated as shown in Table 12.3.

Thus, compared to the normal range of 85–150 pg/ml, Alan G. has a substantially decreased plasma glucagon while the others have high plasma glucagon, Barbara W.'s being particularly elevated. A group of normal individuals should always be used to generate reference data rather than a single normal control since there may be considerable variation even in the normal population. In addition, it may be necessary to choose normal controls of the same sex, age, etc. to have meaningful data for comparison. Control data are normally given as a range (as in this problem) or as the mean ±SD.

The central nervous system is particularly dependent on an adequate supply of glucose. Under conditions of severe hypoglycaemia, as experienced by Alan G., the brain functions abnormally and limb seizures result. By feeding him frequently, an adequate plasma glucose concentration can be maintained to relieve his symptoms.

In the fasting state, hypoglycaemia signals the pancreatic A cells to secrete glucagon whose target, the liver, releases glucose from mobilised glycogen stores. Alan G. appears to be unable to produce enough glucagon to counteract his hypoglycaemia, probably because of a pancreatic A-cell deficiency. However, he is able to respond normally to injected glucagon and so glucagon-replacement therapy (analogous to insulin treatment of diabetic patients) might provide a basis for his treatment. Alan's plasma insulin should be very low because of his hypoglycaemia. The value actually

Table 12.3. Plasma glucagon concentrations calculated from RIA.

Patient	Plasma glucagon (pg/ml)
Alan G.	55
Barbara W.	800
Fred C.	240

reported in the case (Table 12.1) is not quite as low as might have been expected but the reasons for this are not evident.

Barbara W. on the other hand has a massively elevated plasma glucagon concentration which causes the liver to maintain a hyperglycaemic state. Insulin levels rise secondarily in an attempt to counteract this hyperglycaemia. Here the likely cause is a malignant tumour of the pancreatic A cells (glucagonoma). This would have to be treated with chemotherapy, but the prospects for such patients are not good. The skin lesions seen in Barbara W., necrolytic migratory erythema, are a characteristic dermatological feature of hyperglucagonaemia. As a result, this condition, though relatively rare, is often first diagnosed by dermatologists.

Fred C. is asymptomatic. His plasma glucose and insulin are both normal and he also responds normally to injected glucagon. Yet the RIA suggests a significantly elevated plasma glucagon concentration. This conundrum can be explained by considering how the RIA works. It estimates any material that will cross-react with the antiglucagon antibody whether or not it has glucagon activity. Besides normal glucagon, Fred C.'s plasma may contain other proteins that cross-react with the antibody but are devoid of glucagon activity. Several types of such cross-reacting material have been described in asymptomatic individuals, including proglucagon secreted prematurely from the A cells. An alternative explanation is that Fred's glucagon has an inherently low intrinsic activity and so is synthesised in larger quantities to compensate. Fred is particularly illustrative of the dangers inherent in relying on a single 'normal' or asymptomatic control with which to compare patient data.

In an average clinical biochemistry laboratory, the bulk of RIAs are for assay of TSH, human chorionic gonadotropin and luteinising hormone. The assay of 200 samples per week for TSH is typical whereas only one insulin assay per month might be required. Glucagon assays are performed extremely rarely.

Further Questions

1. To what extent would the symptoms of glucose 6-phosphatase deficiency (von Gierke's glycogen storage disorder) and insulinoma resemble those of Alan G.? Devise a protocol for their differential diagnosis.

2. Apart from liver glycogen, what other reserves can the body draw on to maintain the blood glucose concentration? What mechanisms are involved?

3. Glucagon action on hepatocytes does not involve its entry into the cell. Predict the molecular mechanism whereby the hormonal signal is transduced across the plasma membrane. How is the hormonal specificity conferred?

4. What other hormones are produced by the endocrine pancreas? What else does the pancreas produce and how does this explain why early attempts to purify insulin were unsuccessful before Banting and Best showed the way in 1921?

5. What are the advantages and disadvantages of immunological methods in medical diagnosis? Can you distinguish between RIA and ELISA methods?
6. What are the biochemical and physiological consequences of sustained hyperglycaemia?
7. Apart from nervous tissues, what other tissues or cells have an absolute requirement for glucose?
8. One of the ways in which proteins may be radiolabelled is with [^{125}I]iodide. Which residues in the protein would be labelled? What parallel does this have in normal physiology?

Connections

- Use this problem also to review the immune response and the structure of antibody molecules. Distinguish between 'polyclonal' and 'monoclonal' antibodies and describe how they are produced.
- Use this problem to consider the use of radioisotopes in biochemistry and medicine. Think what alternative methods there are to the use of radioisotopes in assay procedures and what their relative advantages are.
- This problem should lead you to consider how the blood glucose level is controlled. Remember the central role of the liver in energy metabolism, especially glycogenolysis and gluconeogenesis, and the antagonistic effects of insulin (generally anabolic) and glucagon, adrenaline, growth hormone and corticosteroids (catabolic). (See also Problems 2 and 17.)
- Outline the synthesis and secretion of the pancreatic peptide hormones. You should know something of the proteolytic processing of prohormones, the role of signal peptides, rough endoplasmic reticulum and the Golgi apparatus. Remember the anatomy of the pancreas and both its endocrine and exocrine functions. (See also Problems 8, 9, 16 and 18.)
- Recall what metabolic fuels are used by the untreated diabetic individual and what the consequences are. Compare the metabolism of a diabetic with that of someone who is starving. You should understand what ketoacidosis is, how it arises, its consequences and how they would be treated. (See also Problems 1, 2 and 10.)
- The action of peptide hormones such as glucagon requires transmission of a signal across the plasma membrane. Review the various components involved in this signalling process (receptors, G proteins, second messengers, protein kinases and phosphatases). (See also Problem 15.)
- Ensure that you understand the statistical terms 'mean', 'standard deviation' and 'normal distribution' and how statistical methods can be used to compare sets of data (for instance patients versus controls).

References

Edwards R (1985) *Immunoassay: An Introduction*. Heinemann, London.

Gosling J P (1990) A decade of development in immunoassay methodology. *Clinical Chemistry* **36**, 1408–27.

Gow I F and Williams B C (1989) Immunoassays for antigens. *Current Opinion in Immunology* **1**, 940–7.

Kemeny D M (1991) *A Practical Guide to ELISA*, pp 1–115. Pergamon Press, Oxford.

Montague W (1983) Glucagon, Chapter 5, pp 86–102. In *Diabetes and the Endocrine Pancreas: A Biochemical Approach*. Croom Helm, London and Canberra.

Vidnes J and Øyasæter S (1977) Glucagon deficiency causing severe neonatal hypoglycaemia in a patient with normal insulin secretion. *Pediatric Research* **11**, 943–9. Original report.

Many of the 'standard' biochemistry and clinical chemistry textbooks have short sections on the principles and applications of immunological methods.

13 NOW WASH YOUR HANDS!

Introduction

Cholera is a disease caused by *Vibrio cholerae*, a Gram-negative bacterium with a single flagellum which is sometimes referred to as the 'comma bacterium' because of its distinctive, curved shape. Cholera is characterised by severe diarrhoea leading to extensive dehydration and loss of electrolytes. Often 10–15 litres of fluid a day may be lost and, if untreated, the disease is fatal in up to 50% of cases. Death results from dehydration, acidosis, loss of K^+, or usually a combination of these. The classical description of the symptoms of 'Asian cholera' provided by the Portuguese physician Garcia da Orta in 1563 is just as apt today:

> The pulse is very weak and can scarcely be felt. The skin is very cold, with some sweat, also cold. The patient complains of great heat and a burning thirst. The eyes are much sunken and he cannot sleep. There is continued retching and diarrhoea until he is unable to discharge anything. There is also cramp in the legs . . . It generally proves fatal within 24 hours.

Cholera is spread by faecal contamination of drinking water. The bacteria multiply rapidly in the small and large intestines and, after a short incubation period of 1–6 days, the diarrhoea commences. Eradication is, in principle, a simple matter of preventing faecal contamination of water supplies. This was first recognised, long before the bacteriological cause of the disease was realised, in the brilliant epidemiological investigations of John Snow during the London cholera epidemics of 1853 and 1854. Reputedly, his removal of the Broad Street pump handle led to the waning of the Soho cholera outbreak. Today, cholera probably accounts for the deaths of about four million children every year and has recently made a dramatic reappearance in South America. In principle, however, the disease is readily preventable and is treatable with antibiotics.

The disease is caused by the protein toxin produced by the bacteria. This toxin is an enzyme that causes the intracellular concentration of the second messenger, cyclic AMP (cAMP), to rise. In the intestinal cells this produces uncontrollable secretion of water and Cl^- into the lumen and an inhibition of Na^+ absorption. Although the major sodium transport system is shut down there is, however, an alternative route, the so-called 'co-transport' system. This system only operates when *both* Na^+ and glucose are present in the gut lumen and it does not appear to be affected by cholera toxin.

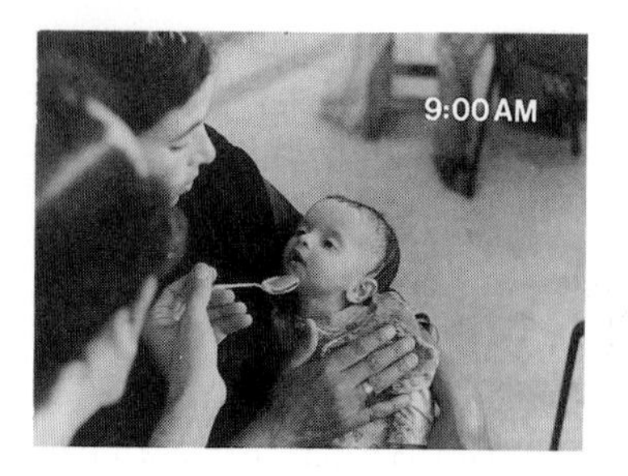

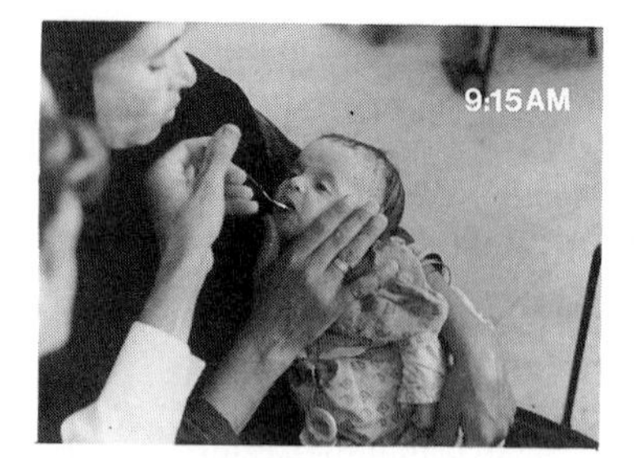

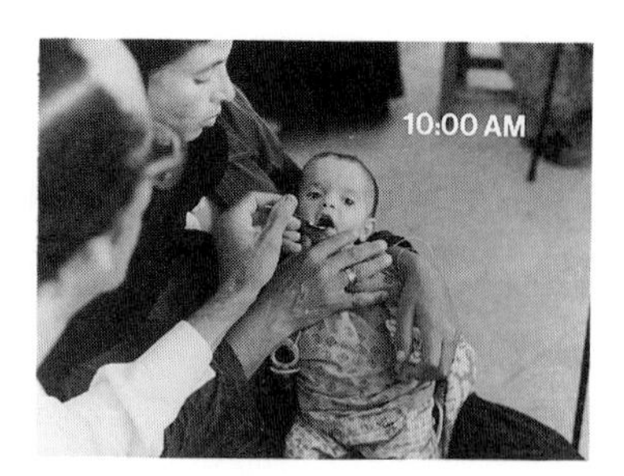

Figure 13.1. Time course of rehydration in a baby. Photographs kindly supplied by Dr Mark Robbins (John Snow Inc., Boston).

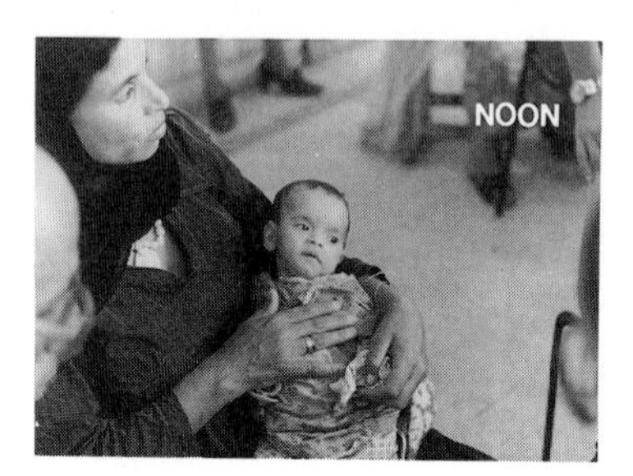

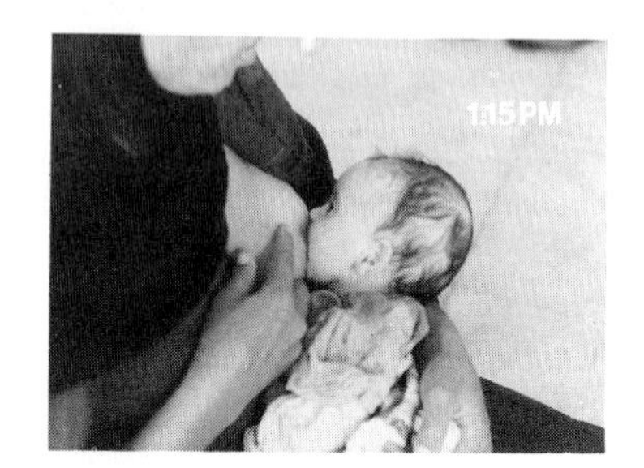

The Problem

A young African baby, Abdul N., was found to be suffering from severe diarrhoea, producing large amounts of watery stools. He exhibited the typical signs of dehydration, including limpness, sunken eyes and flattening of the soft spot at the top of the skull. The baby was administered an oral rehydration solution containing glucose, sodium chloride, potassium chloride and sodium bicarbonate. Within 4 hours the baby had shown some signs of recovery from the dehydration and was able to take in breast milk from his mother. The mother continued the rehydration therapy with small but frequent feedings of about half a litre of solution a day for six days until the diarrhoea had run its course. The dramatic and rapid effects of oral rehydration are shown in the series of pictures in Figure 13.1.

Questions

1. What are the basic biochemical principles underlying the rehydration therapy?
2. Remembering that cholera is a disease of developing countries, explain why the oral rehydration therapy described above is preferable to vaccination, treatment with antibiotics such as tetracyclines, or intravenous rehydration therapy with glucose–saline?
3. Some formulations for oral rehydration therapy contain sodium bicarbonate. Others contain lactate, which is more stable than bicarbonate, or citrate (see Figure 13.2). What is the function of these components?

Figure 13.2. Example of a widely available oral rehydration pack (reproduced with permission of Rhone-Poulenc Rorer Ltd).

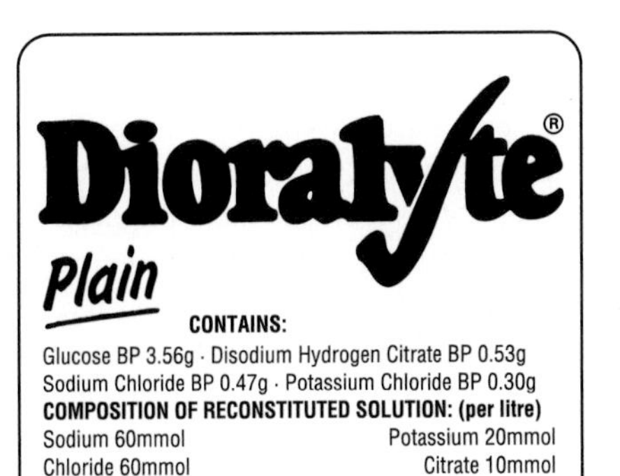

Dioralyte is the modern way to treat the loss of body fluid and salts that occurs in acute diarrhoea and other conditions

HOW TO USE: (Dioralyte makes up into a colourless solution).

1. Dissolve contents of one Dioralyte sachet in 200ml (7floz) of fresh drinking water. **For infants,** or where fresh water is not available, use freshly boiled and cooled water.
2. Dioralyte should **only** be made up with water **and** to the volume stated. A greater or lesser dilution may result in electrolyte imbalance

INFANTS UNDER 12 MONTHS: Take **ONLY** if recommended by a doctor.

DOSAGE: You **MUST** consult the leaflet provided before taking this medicine.
If the illness is not improving within 36 hours consult your doctor.
Made up Diarolyte solution may be used for up to 24 hours **if kept refrigerated** otherwise discard within one hour.

Batch No: Use Before:

4. Glucose is present in the oral rehydration mixture at a concentration of 0.11 M. If glucose is beneficial, why cannot a much higher concentration of glucose be used, both to assist the co-transporter and to provide additional energy for the patient who is likely to be malnourished?
5. A more recent variant of the oral rehydration therapy undergoing trials involves the substitution of starches (e.g. rice powder) for glucose. What advantages might this new treatment offer?
6. How does cholera toxin bring about a rise in the intracellular concentration of cAMP?

Commentary

A primary concern in treatment of the patient is to replace the massive amounts of water lost and hence prevent further dehydration. The therapy depends upon the Na^+–glucose co-transport system (symport) in the intestinal villus cells (Figure 13.3).

Many cell-transport systems are coupled to the flow of an ion down its electrochemical gradient. This includes the transport of sugars, amino acids and some neurotransmitters (e.g. catecholamines, γ-aminobutyrate). In the gut the uptake of glucose is accompanied by the uptake of Na^+, both of which must bind to the transport protein for it to operate. The sodium ions are subsequently pumped out of the cell on the serosal side into blood through the action of the Na^+K^+-ATPase. Although the normal function of the intestinal transporter is to allow uptake of glucose for subsequent metabolism, the basis of the rehydration therapy is to make use of the ability of the transporter to take up sodium ions in the presence of glucose. This reverses the sodium loss occurring in the cholera patient and hence allows the replacement of water which follows the Na^+ by an osmotic effect and reverses the dehydration.

Remember that economics, as well as efficacy, are important factors in treatment in a developing country. Cholera vaccines are of limited usefulness. In field trials conducted in areas of endemic cholera, vaccines have been only about 50% effective in reducing the incidence of clinical illness. They also afford protection for only about six months to one year and they do not prevent transmission of infection. Antibiotics are expensive and may even make the situation worse by killing beneficial intestinal bacteria! Intravenous rehydration therapy, which was standard practice before 1978, requires sterile solutions and needles, as well as trained personnel. The oral rehydration therapy, however, is cheap and remarkably effective. The solution is made up from inexpensive chemicals, which can be supplied in pre-weighed sachets, and made up with boiled water. Large numbers of individuals, including children, may be treated by relatively inexperienced personnel or by the child's parents. The success of the therapy relies on the sodium–glucose co-transport system remaining operational in the patient, which is normally the case.

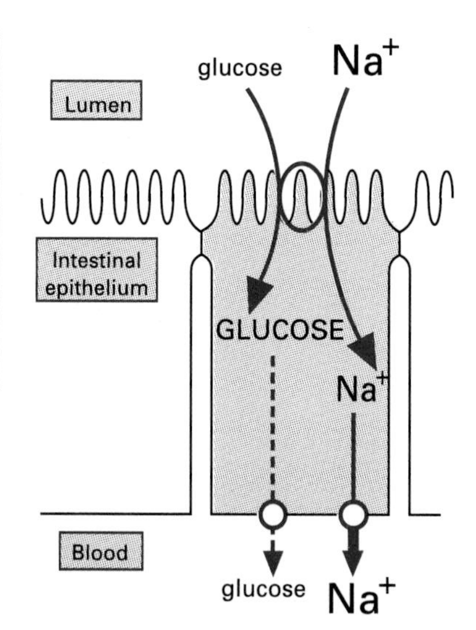

Figure 13.3. Outline of villar uptake of glucose and sodium. Reproduced from Finean *et al.* (1979) *Membranes and their Cellular Function*, Blackwell Scientific Publications, Oxford, with permission.

The optimal formulation of oral rehydration fluids is still a matter of debate. The roles of sodium and glucose have been explained above. The bicarbonate (or lactate) originally used in rehydration mixtures helped to prevent the acidosis associated with electrolyte loss in the diarrhoea. Citrate has better buffering power than lactate and helps to return the blood pH to normal quickly. The presence also of potassium ensures the normal functioning of cells. The problem with increasing any of the solutes present in the oral rehydration mixture is that the solution would then be more concentrated than blood and would therefore tend to exacerbate the dehydration which is just the opposite of what is wanted.

A polymer of glucose such as starch has the same osmotic effect as a single molecule of glucose. Although the starch is broken down at the intestinal surface, the individual glucose molecules are immediately absorbed and so do not increase the osmolarity in the intestinal lumen. However, the increased number of glucose molecules entering the villus cells together with the sodium speeds up the return of water and ions to the blood hence reducing the extent and duration of the diarrhoea. The starch also helps to increase the bulk of the stools excreted. Another modification of oral rehydration therapy is to include amino acids, typically alanine, as well as glucose, in the mixture. Alanine, like glucose, is absorbed by a Na^+-dependent symport mechanism and its presence enhances the uptake of sodium and hence water.

The oral rehydration procedure using starches in place of glucose also has applications in the post-operative management of patients who have had multiple bowel resections since they tend to lose large quantities of fluid. Rice powder combined with oral rehydration therapy can dramatically reduce this loss and the stool output in such patients.

Many peptide hormones are coupled via guanyl-nucleotide binding proteins (G proteins) to the activation of adenylate cyclase. In their active form the G proteins have GTP bound and subsequent hydrolysis of the GTP by the intrinsic GTPase activity of the G protein switches off the hormone-activated cascade. Cholera toxin increases intracellular cAMP by 'ADP-ribosylating' the α-subunit of a G protein in the intestinal cell membrane which inhibits its GTPase activity thereby causing irreversible activation of adenylate cyclase. The cAMP produced activates a protein kinase which probably regulates electrolyte secretion, hence controlling water excretion. The toxin has two types of subunit: five *B subunits* arranged in a ring bind to a ganglioside molecule on the cell surface thus enabling the catalytically active *A subunit* to penetrate into the cytosol where it covalently modifies an arginine residue on the G protein α-subunit by using NAD^+ as substrate:

$$\alpha\text{-subunit} + NAD^+ \rightarrow \text{ADP-ribosyl-}\alpha\text{-subunit} + \text{nicotinamide} + H^+$$

The intestinal epithelial cells are eventually shed into the gut and are replaced by new cells which will then function normally as long as the cholera infection has disappeared.

Cholera is a serious disease, usually of the poor; tourists rarely contract it. However, 'traveller's diarrhoea' (often evocatively referred to as 'Delhi belly', 'Montezuma's revenge', etc.) is fairly common. A major cause is probably a heat-labile enterotoxin

(LT) secreted by *Escherichia coli*. LT is closely related to cholera toxin and acts by a similar mechanism. Heat-stable enterotoxins are also secreted into the intestine by certain strains of *E. coli*. These toxins are small, cysteine-rich peptides (hence their heat stability) which induce diarrhoea by stimulation of a receptor linked to guanylate cyclase. The resultant increase in cyclic GMP inhibits salt absorption and stimulates chloride secretion leading to the diarrhoea and dehydration typical of enterotoxin activity.

Two potentially lethal bacterial toxins, botulinum and tetanus toxins, exert their toxic effects through an entirely different enzyme mechanism. These toxins enter nerve cells where the active subunits, which are zinc-containing metalloproteinases, cleave specifically one or more proteins involved in neurotransmitter release.

Further Questions

1. Review the structure and mechanism of ion-channel proteins. What are the differences between 'facilitated diffusion' and 'active transport'? How can they be distinguished experimentally? How does the antibiotic valinomycin work?
2. G proteins form part of many signal transduction cascades in cells, and the ultimate effect of hormones and other stimuli may be inhibitory rather than stimulatory. Find other examples of G protein systems and make sure you know how the cascades result in amplification of the original signal.
3. Diphtheria toxin and pertussis toxin also act on guanyl-nucleotide binding proteins. Which proteins and with what effects?
4. How might you measure the levels of cAMP in a tissue extract?
5. What other 'second messengers' do you know of, apart from cAMP?
6. The cholera toxin receptor is a ganglioside. What type of cell-surface molecule is this?

Connections

- This question is really concerned with the cell membrane, a complex structure whose properties we still do not appreciate fully. Do you understand the basic lipid–protein structure of cell membranes and can you explain the need to have transport arrangements for molecules that need to cross the membrane?
- The most important connection in this problem is between hygiene and the spread of disease-causing bacteria – hence 'Now Wash Your Hands!' It has been said that many more lives have been saved by improved sanitation and good water-purification systems than by all the efforts of the pharmaceutical companies.
- Use this problem to review the structure of NAD^+ and related coenzymes.
- Recall the roles of adenylate cyclase and cAMP. Where is the cyclic structure in cAMP? (See also Problem 15.)

- Do you understand the difference between 'respiratory' and 'metabolic' acidosis? Why does uncontrolled diabetes mellitus lead to acidosis? (See also Problems 10 and 18.)
- Cholera is closely associated with conditions of malnutrition. Do you understand the difference between marasmus and kwashiorkor? (See also Problem 2.)
- Breast milk is good! It comes sterile and has the correct nutritional composition for human infants. It also has some antibacterial properties and can confer immunity to some diseases. Can you explain why?
- What are Gram-negative bacteria and why are they so called? This will give you an opportunity to look up some basic bacteriology.
- Most bacterial infections can be treated with antibiotics. Check that you understand the basic principles involved; illustrate these by considering penicillin, chloramphenicol and the sulphonamides.

References

Behrens R H (1991) Cholera. *British Medical Journal* **302**, 1033–4.

D'Cruz I A (1991) Garcia da Orta in Goa: pioneering tropical medicine. *British Medical Journal* **303**, 1593–4.

Hirschhorn N and Greenough W B (1991) Progress in oral rehydration therapy. *Scientific American* **264**(5), 16–22.

Huttner, W B (1993) Snappy exocytoxins. *Nature* **365**, 104–5. *News and Views* article.

Levine M M (1991) South America: the return of cholera. *Lancet* **338**, 45–6.

Moss J and Vaughan M (1989) Guanine-nucleotide binding proteins (G proteins) in activation of adenylyl cyclase: lessons learned from cholera and travelers diarrhea. *Journal of Laboratory and Clinical Medicine* **113**, 258–68.

Rietschel E T and Brade H (1992) Bacterial endotoxins. *Scientific American* **267**(2), 26–33.

Robbins A and Freeman P (1988) Obstacles to developing vaccines for the Third World. *Scientific American* **259**(5), 90–5.

Vandenbroucke J P, Eelkman Rooda H M and Beukers H (1991) Who made John Snow a hero? *American Journal of Epidemiology* **133**, 967–73.

Watson A J M (1992) Diarrhoea. *British Medical Journal* **304**, 1302–4.

14 BLADDER CANCER

Introduction

Cancers account for about 20% of deaths in the UK and USA (Figure 14.1). The development of cancer (oncogenesis) is multifactorial, i.e. it requires several deleterious events to produce a tumour. Among these is exposure to environmental carcinogenic agents. This environmental aetiology has become increasingly obvious ever since Sir Percival Pott recognised that the high incidence of cancer of the scrotum in the boy chimney sweeps of 18th Century London was due to exposure to soot. We now recognise a causal connection between tobacco smoking and lung cancer, asbestos and mesothelioma, ultraviolet light and skin cancer, and so on. However, what is still somewhat of a mystery is why exposure to a particular carcinogen results in a very characteristic tissue-specific pattern of cancer. This problem is concerned with working out why a particular carcinogen causes bladder cancer.

Figure 14.1. Major causes of death (a) and cancer mortality (b) in the USA.

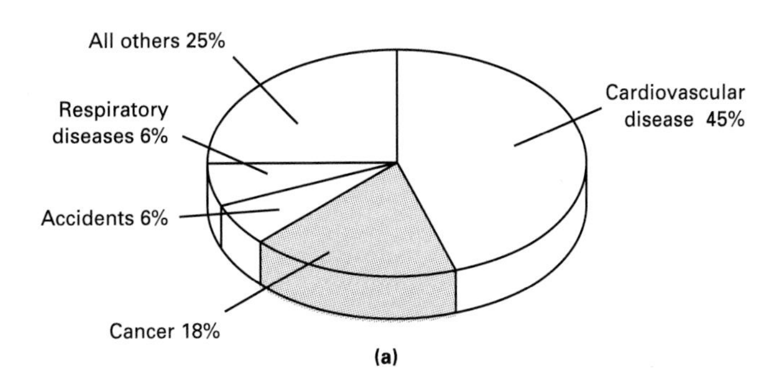

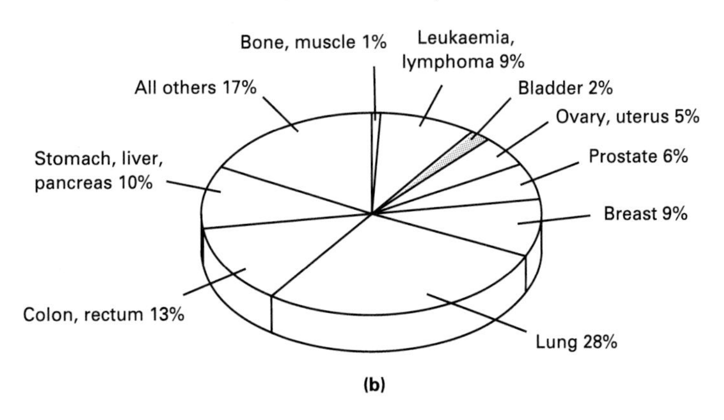

The Problem

Some time ago, it was observed that there was a very high incidence of bladder carcinoma among workers exposed to 2-naphthylamine (β-naphthylamine):

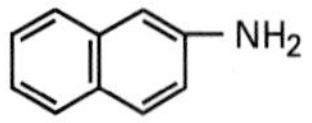

2-Naphthylamine

This aromatic amine was widely used in the preparation of synthetic dyestuffs and as an antioxidant in the synthetic rubber industry, for instance in making car tyres. In view of its carcinogenic potential, occupational exposure to this chemical is now severely restricted, but it also occurs in tobacco smoke and may be responsible for the increased incidence of bladder carcinoma observed among smokers.

As part of an experimental study to elucidate the basis of the tissue-selective action of 2-naphthylamine, a series of metabolic experiments was performed:

1. Incubation of human liver microsomes with 2-naphthylamine plus NADPH and UDP-glucuronic acid resulted in the formation of *N*-hydroxy-2-naphthylamine-β-1-glucuronyl pyranoside (NHNGP):

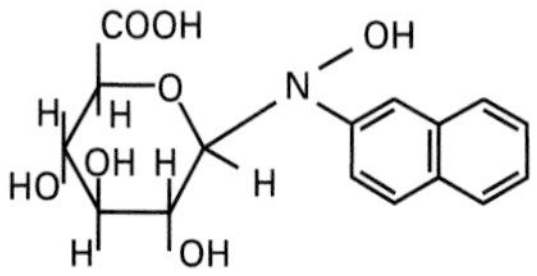

   NHNGP

2. Human bladder tissue incubated with 2-naphthylamine formed *N*-acetyl-2-naphthylamine:

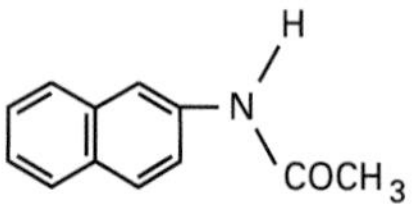

   N-Acetyl-2-naphthylamine

3. NHNGP, 2-naphthylamine and *N*-acetyl-2-naphthylamine did not bind strongly to DNA nor did these compounds stimulate DNA repair in hepatocytes.

4. In the pH range of 5.5–6.5, the *N*-glycosidic bond in NHNGP was readily hydrolysed. One of the products was strongly reactive with DNA.

Questions

1. Why were the experimenters interested in whether or not 2-naphthylamine or its metabolites bound to DNA or stimulated DNA repair?
2. Why were microsomes, NADPH and UDP-glucuronic acid included in experiment 1? What types of enzymic reactions might the liver use to metabolise 2-naphthylamine?
3. What is the significance of the pH range chosen in Experiment 4?
4. Considering all of the observations in Experiments 1–4, why do you think that exposure to 2-naphthylamine results in bladder, but not liver, carcinoma?

Commentary

Chemical carcinogenesis is associated with somatic cell mutations ranging from point mutations to large-scale chromosome aberrations. There is a good correlation between the ability of a compound to cause mutations and its oncogenicity. Many carcinogens bind to DNA and react with it, producing mutations. Fortunately, most of the mutations which DNA experiences on a daily basis from a wide variety of environmental agents are recognised and reversed by various DNA repair systems operating in somatic cells. The few mutations which escape the vigilance of the repair systems may contribute to a malignancy.

The inability of 2-naphthylamine to bind to DNA and to stimulate DNA repair systems suggests that 2-naphthylamine itself is not a mutagen and therefore not likely to be a carcinogen. However, many environmental agents are so-called pre-carcinogens, compounds which are themselves harmless but which are metabolised to highly-reactive carcinogenic derivatives.

Although bladder tissue metabolises 2-naphthylamine to *N*-acetyl-2-naphthylamine, this metabolite is not the active carcinogen since it neither bound to DNA nor stimulated DNA repair. The origin of the active carcinogen must lie elsewhere. In fact the metabolic route to the carcinogen starts in the liver with enzyme systems for the metabolism of xenobiotics (foreign compounds). These 'detoxification' systems result in conversion of xenobiotics to more soluble and easily excreted derivatives which are usually less toxic. Metabolism occurs in two stages. Stage one involves hydroxylation, where molecular oxygen is the source of the oxygen atom for the hydroxyl group; the other oxygen atom is reduced to water. Hydroxylation involves a specialised electron-transport system bound to the membranes of the smooth endoplasmic reticulum (microsomes). NADPH provides the reducing power via a specialised cytochrome system, cytochrome P_{450}:

$$\text{2-naphthylamine} + O_2 + \text{NADPH} + H^+ \rightarrow \\ \text{N-hydroxy-2-naphthylamine} + H_2O + \text{NADP}^+$$

N-hydroxy-2-naphthylamine is rapidly metabolised further by the stage two conjugation system, whose enzymes are also bound to the microsomal membranes in close proximity to the hydroxylation system. *N*-hydroxy-2-naphthylamine does not accumulate as such but rapidly has a polar glucuronate unit attached to it to form NHNGP. This reaction is catalysed by a glucuronyl transferase using the activated substrate, UDP-glucuronic acid.

Even NHNGP is not the active carcinogen but it follows one of the usual excretory routes for conjugated xenobiotics via the kidneys into the urinary bladder. Here the problems start. Urine is usually somewhat acidic (Appendix 2) and provides just the right conditions for the hydrolysis of NHNGP to *N*-hydroxy-2-naphthylamine. This accumulates in the bladder and, under acidic conditions, can be further converted to a variety of highly-reactive derivatives. These reactive derivatives readily bind to the DNA of bladder epithelial cells, setting in train the eventual bladder carcinoma.

In a sense we can regard the bladder-specific detrimental effects of 2-naphthylamine as an unfortunate accident of biochemistry with catastrophic consequences. What started out as a straightforward attempt to detoxify and excrete the xenobiotic, 2-naphthylamine, inadvertently exposes bladder tissue to a derivative far more dangerous than the original compound.

Further Questions

1. Bearing in mind the information in this problem, how would you design a rapid screening test for potential carcinogens like 2-naphthylamine?
2. Why is there an increased risk of cancer in conditions such as xeroderma pigmentosum, Fanconi's anaemia and Bloom's syndrome?
3. Many cancers involve chromosome aberrations, including translocations. How would you screen cells for specific chromosome aberrations? What is the Philadelphia chromosome?
4. Microsomal enzyme systems are not just concerned with the transformation of xenobiotics, they are also involved in the metabolism of endogenous compounds. What roles do they have in normal metabolism?
5. The acidity of urine contributed to the bladder-specific carcinogenicity of 2-naphthylamine. Why is urine normally acidic? If you needed to reduce the acidity of a patient's urine, what would you do?
6. Death of cancer patients is rarely due to the primary tumour; rather it results from dissemination of tumour cells to other body sites (metastasis). From what you know about the structure of epithelia, what phenotypic changes would be necessary for malignant epithelial cells to metastasise and establish tumours elsewhere?

Connections

- The liver is not just concerned with detoxifying xenobiotics; potentially toxic compounds are also produced endogenously, e.g. bilirubin. Review the process of haem degradation, including haemolysis, the role of the spleen and gut bacteria, jaundice, and bilirubin status as an indicator of liver function. (See also Problems 6 and 17.)
- List the different ways DNA can be mutated and later 'repaired' in the cell. Ensure you understand the connection between sunlight and skin cancer.
- During DNA replication the newly-synthesised DNA strand is 'proof-read' for copying errors. Check that you know how this is done.
- UDP-glucuronate is an example of an 'activated' nucleotide. List other examples (for example, recall the biosynthesis of glycogen and phospholipids, and the active methyl cycle). Recall why such compounds are necessary.

- Look up the Ames test for screening carcinogens and check that you understand its basis.
- Oncogenes were identified in oncogenic retroviruses and this led to the discovery of cellular oncogenes. Remind yourself what an oncogene is and the difference between viral and cellular oncogenes. Recall that there are also tumour suppressor genes; look up some examples and see how they are thought to act.

References

Egan S E and Weinberg R A (1993) The pathway to signal achievement *Nature* **365**, 781-3. *News and Views* article on signal transduction and oncogene products.

Feldman M and Eisenbach L (1988) What makes a tumor cell metastatic? *Scientific American* **260**(5), 40–7.

Garner C (1992) Molecular potential. *Nature* **360**, 207–8. *News and Views* article on new methods for detecting carcinogens through DNA adducts in biopsy samples.

Goldberg A M and Frazier J M (1989) Alternatives to animals in toxicity testing. *Scientific American* **261**(2), 16–22.

Josephy P D (1989) New developments in the Ames assay: high-sensitivity detection of mutagenic aryl amines. *Bioessays* **11**, 108–12.

Radman M and Wagner R (1988) The high fidelity of DNA duplication. *Scientific American* **259**(2), 24–30.

Varmus H E and Weinberg R A (1992) *Genes and the Biology of Cancer.* Scientific American Library, New York.

Weinberg R A (1988) Finding the anti-oncogene. *Scientific American* **259**(3), 34–41.

15 SIGNAL FAILURE

Introduction

The maintenance of homeostasis in multicellular organisms requires a sophisticated mechanism of cell–cell communication via circulating chemical messengers such as hormones, growth factors and cytokines. Some of these factors are relatively hydrophobic (e.g. steroid hormones) and are able to enter their target cells to exert their effects. Many signal molecules, however, are too large or too hydrophilic (e.g. peptide hormones) and so must exert their effects at the cell surface. This requires a system of highly-specific receptors at the surface and a membrane signal transduction system to transfer the 'message' across the membrane and propagate it within the cell. Several different classes of membrane receptor systems operate at the cell surface and these fall into a number of receptor 'superfamilies' as illustrated in Figure 15.1.

Among signal molecules working via signal transduction pathways are growth regulators whose role is to control cell proliferation and differentiation, e.g. epidermal growth factor. Some cancers are characterised by breakdown of these controls over cell proliferation. An integral step in tumorigenesis appears to be mutation of certain key regulatory genes (the cellular or proto-oncogenes). Many of these genes have been shown to specify components of transmembrane signalling pathways. Apart from cancers there are other disorders which feature abnormalities in the perception and transduction of chemical signals as in the problem which follows.

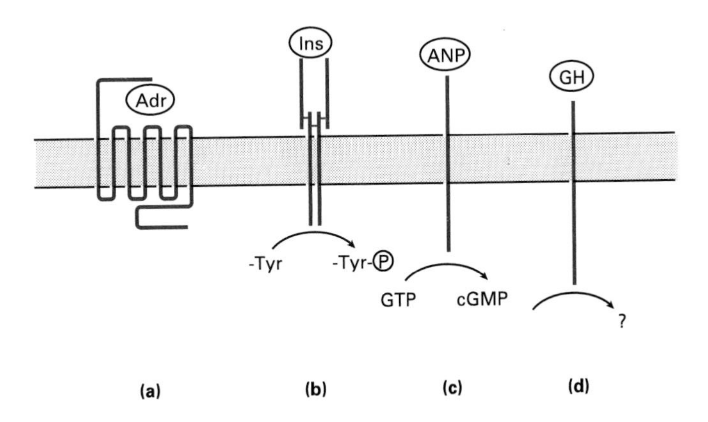

Figure 15.1. Examples of the topology of transmembrane receptors for hormones and growth factors. Predicted structures are shown for (a) G protein-linked receptors, (Adr = adrenaline), (b) insulin (Ins)/growth factor receptors, (c) guanylate cyclase-linked receptors, e.g. for the hormone atrial natriuretic peptide (ANP), (d) the growth hormone (GH), prolactin, interleukin family.

The Problem

Since the age of 4, Mary H., now aged 13, had suffered convulsions and muscle cramps which responded to treatment with calcium salts and vitamin D. She had a short and stocky build with a number of skeletal abnormalities, especially metacarpal and metatarsal shortening, and poor tooth development. There was evidence of calcification in subcutaneous tissues. On X-ray examination, however, the bone density was found to be normal, indicating proper mineralisation of bone. Mary also had a marked reduction in her sense of smell.

Her serum calcium and phosphate levels, before and after administration of 1,25-dihydroxy vitamin D_3 (Figure 15.2), were as indicated in Table 15.1; there was little calcium in the urine. The production of parathyroid hormone (PTH) was then measured in the patient before and after an infusion of Ca^{2+} to raise calcium levels to the normal range. The results are shown in Table 15.2. The patient's urinary excretion of cyclic AMP in response to infused PTH was less than 10% of normal. The specific binding of radiolabelled PTH to renal membranes was in the normal range. Mary's plasma levels of thyroid stimulating hormone (TSH) were also above normal, whereas her thyroid hormone levels were at the lower end of the normal range.

Figure 15.2. Structure of 1,25-dihydroxy vitamin D_3.

CH3 CH3 CH3 CH3 25 OH OH H 1 H HO

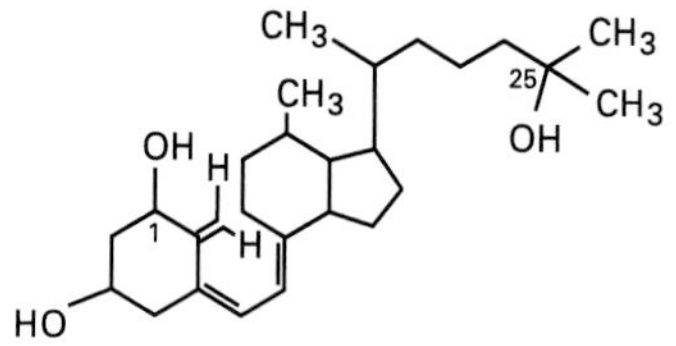

Table 15.1. Effects of 1,25-dihydroxyvitamin D_3 on serum calcium and phosphate levels.

	Normal range	Patient	Patient after administration of 1,25-$(OH)_2$ vitamin D_3
Calcium (mM)	2.1–2.7	1.1	2.6
Phosphate (mM)	0.7–1.4	2.8	1.3

Table 15.2. Effects of calcium infusion on PTH and calcium levels in the patient.

	Normal range	Patient	Patient after Ca^{2+} infusion
Calcium (mM)	2.1–2.7	**1.8**	2.9
PTH (ng/ml)	<0.5	**3.1**	undetectable

Questions

1. What are the normal physiological functions of PTH?
2. What method could be used routinely in a clinical biochemistry laboratory to measure levels of PTH in samples from patients?
3. From the data given above, can you deduce one signal transduction mechanism through which PTH operates?
4. Why might *both* PTH and TSH fail to exhibit their normal physiological responses?
5. When membranes prepared from a sample of the patient's erythrocytes were treated with cholera toxin and radiolabelled NAD^+, a single polypeptide of M_r approximately 45 000 was radiolabelled but only to 40% of the level in membranes from control subjects. What does the experiment with cholera toxin indicate about the possible molecular defect in the patient?
6. How might the proposed defect lead to the clinical picture presented?

Commentary

PTH, together with 1,25-$(OH)_2$ vitamin D_3 and calcitonin, provide homeostatic control of blood Ca^{2+} levels. PTH is secreted in response to hypocalcaemia and raises the serum Ca^{2+} concentration to approximately 2.7 mM, when hormone production stops. PTH increases Ca^{2+} reabsorption by the renal tubule, stimulates dissolution of bone mineral and aids in the absorption of Ca^{2+} by the gut. PTH also acts to lower serum inorganic phosphate levels and increases renal production of 1,25-$(OH)_2$ vitamin D_3.

PTH exerts its actions principally by stimulating the production of cyclic AMP in bone and kidney. The latter effect is responsible for the appearance of cAMP in the urine of the patient in response to PTH. PTH may also operate through the phosphoinositide signalling system in bone cells. Mary has a defect in cell signalling as a result of a deficiency in a G protein (guanyl-nucleotide binding protein). This particular condition is one form (type 1a) of *pseudohypoparathyroidism* often referred to as 'Albright's hereditary osteodystrophy', an inherited disorder of cAMP production resulting in overproduction of PTH but with symptoms of PTH deficiency because of the failure to respond to the hormone adequately. This condition is therefore analogous to type 2 (maturity onset) diabetes in which there is also a signalling defect: a failure to respond to insulin (insulin resistance). Pseudohypoparathyroidism was first described by Albright in 1942 and represents the first description of target organ hormone resistance as a cause of decreased hormone responsiveness.

A suitable procedure for measuring PTH levels would be a radioimmunoassay. The principle of the assay is based on competition between radiolabelled PTH and the unlabelled PTH present in a blood sample for binding to an antibody specific for PTH. This type of assay is explained in more detail in Problem 12.

That *both* PTH and TSH fail to elicit their normal physiological responses suggests that there is not a defect in the PTH (or TSH) receptors themselves. The receptor-binding data with $[^{125}I]$PTH also confirm this. Rather, the defect must lie in the coupling of these receptors to their common effector system, adenylate cyclase. Patients with hypoparathyroidism type 1a may therefore present with a number of other endocrine disorders in addition to hypoparathyroidism, of which primary hypothyroidism is the most common. Other patients have been reported who show resistance to gonadotropins and to glucagon. The integral membrane receptors for many polypeptide hormones (and for catecholamines) exhibit similar structural features, especially the presence of seven transmembrane spanning domains (Figure 15.1). These receptors couple to effector elements (adenylate cyclase, phospholipase C, ion channels etc.) via members of the family of G proteins, of which more than 20 are now known to occur. Some examples of G proteins and their signalling pathways are provided in Table 15.3. The G proteins are heterotrimeric, being composed of α, β and γ subunits. The α subunit, which interacts with the effector system directly, is a GTPase; hydrolysis of GTP terminates the signal transduction process. The signal transduction cascade is illustrated in Figure 15.3. Note that adenylate cyclase consists of two transmembrane domains each of which spans the membrane six times and which are separated by a large cytoplasmic domain. This topology resembles a larger superfamily of transporter molecules which includes the cystic fibrosis gene product and the multi-drug resistance protein. However, no specific transport functions have been associated with adenylate cyclase.

Table 15.3. G proteins and their principal cellular signalling pathways.

Class	Effector	Second messenger
G_s	Adenylate cyclase	↑ cAMP
G_i family	Adenylate cyclase	↓ cAMP
G_q	Phospholipase C	↑ Inositol *tris*phosphate, diacylglycerol, Ca^{2+}
G_t (transducin)	cGMP phosphodiesterase	↓ cGMP

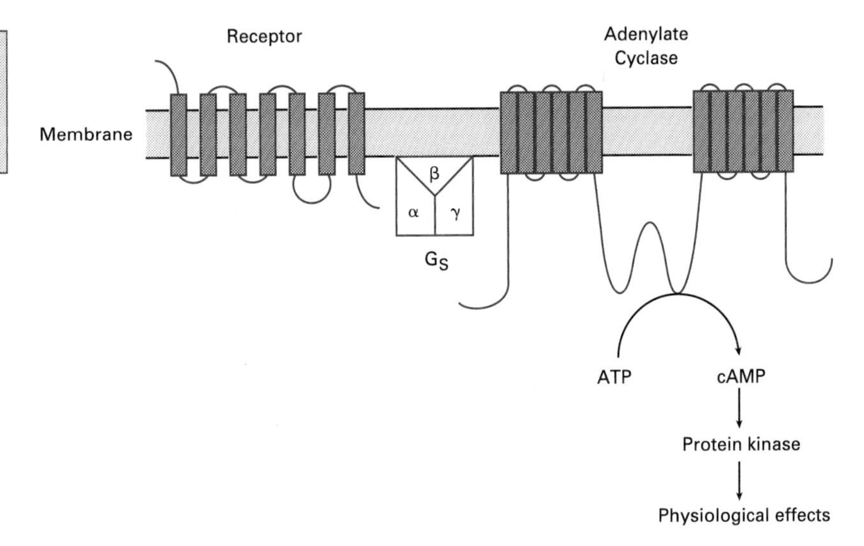

Figure 15.3. Typical G protein signal cascade system involving cAMP.

Another syndrome that involves mutations in a G protein gene and hence abnormal signal transduction is the McCune–Albright syndrome. Some examples of juvenile gigantism may also involve altered G protein and adenylate cyclase activity. It has been suggested that the Tegernsee giant, who died in 1876, may have had just such a defect.

Cholera toxin is an enzyme which can catalyse the transfer of ADP-ribose from NAD^+ to a target protein. This post-translational modification (ADP-ribosylation) occurs on a specific arginine side-chain of the α-subunit of a particular G protein, G_s (M_r approximately 45 000), that stimulates adenylate cyclase. ADP-ribosylation inactivates the GTPase activity of the G protein by obliterating an essential positive charge in the active site which is needed for hydrolysis. The covalent modification leaves the cyclase permanently activated. There is clearly a reduced concentration of G_s protein in the patient resulting in the decreased responsiveness to PTH and TSH. Not all hormones that activate adenylate cyclase are equally affected since there is a family of G proteins, each member of which is expressed to different extents in different tissues.

The abnormal handling of Ca^{2+} suggests a problem in Ca^{2+} homeostasis but the proper mineralisation of bone rules out rickets. Simple vitamin D deficiency is ruled out by the co-existent hyperphosphataemia. The convulsions and muscle cramps are a consequence of the lowered serum calcium levels. The reduction in the sense of smell is a result of the G protein deficiency; olfaction is also mediated through membrane receptor proteins coupled, via a G_S protein, to adenylate cyclase. The short stature and shortening of metacarpals and metatarsals are typical of patients with pseudohypoparathyroidism type 1a and may be due to premature skeletal maturation. The relationship between these phenotypic features, as well as the calcification of subcutaneous tissues, and the G protein deficiency remains unclear. Such data as are available suggest normal growth hormone secretion in these patients.

Further Questions

1. How would you expect this patient to respond to the intravenous administration of thyrotropin releasing hormone (TRH) in comparison with controls? How is the circulating concentration of TSH normally regulated?
2. What treatment might you recommend to a patient with hypoparathyroidism type 1a?
3. Although the cholera toxin experiment indicates a possible deficiency in G_S, reduced labelling by cholera toxin could be due to other reasons, for example modification of G_S or alterations in the association of α_s with the $\beta\gamma$ subunits. What further tests, especially immunological, might you perform to support the hypothesis of a reduction in the levels of G_S?
4. What is the molecular mechanism of action of the active form of vitamin D (1,25-dihydroxy vitamin D_3)?

Connections

- Use this problem to review the actions of other related bacterial toxins, e.g. diphtheria toxin and pertussis (whooping cough) toxin. (See also Problem 13.)
- This problem should lead you to consider the role of calcium in cell signalling and the compartmentalisation of Ca^{2+} in the cell.
- Compare signalling mediated by peptide hormones with visual transduction mechanisms. What features does the β-adrenergic receptor have in common with the visual pigment rhodopsin?
- The mechanism of action of vitamin D should lead you to recall how other steroid hormones function. (See also Problems 7 and 22.)
- For another example of a cell signalling defect refer to Problem 18.
- For further consideration of problems of proper bone mineralisation consult Problem 8.

- Some patients with pseudohypoparathyroidism also show symptoms of Parkinsonism such as tremor, rigidity and stooped posture which is thought to be due to calcification of the basal ganglia in the brain. Refer to Problem 21. Ensure you understand what Parkinsonism is and use this knowledge as a basis for revising mechanisms of neurotransmitter action.

References

DeLuca H F (1988) The vitamin D story: a collaborative effort of basic science and clinical medicine. *FASEB Journal* **2**, 224–36.

Egan S E and Weinberg R A (1993) The pathway to signal achievement *Nature* **365**, 781–3. *News and Views* article on signal transduction and the *ras* protein.

Hosking D J and Kerr D (1988) Mechanisms of parathyroid hormone resistance in pseudohypoparathyroidism. *Clinical Science* **74**, 561–6.

Lefkowitz R J (1993) Turned on to ill effect. *Nature* **365**, 603–4. *News and Views* article on G protein-coupled receptors in disease.

Levine M A (1991) The McCune–Albright syndrome: The whys and wherefores of abnormal signal transduction. *New England Journal of Medicine* **325**, 1738–40.

Linder M E and Gilman A G (1992) G-proteins. *Scientific American* **267**(1), 36–43.

Michell R H (1987) How do receptors at the cell surface send signals to the cell interior? *British Medical Journal* **295**, 1320–3.

Rasmussen H (1986) Mechanisms of disease: The calcium messenger system. *New England Journal of Medicine* **314**, 1094–101.

Reichel H, Koeffler H P and Norman A W (1989) The role of the vitamin D endocrine system in health and disease. *New England Journal of Medicine* **320**, 980–91.

Schwindinger W F, Francomano C A, Levine M A and McCusick V A (1991) DNA light on the Tegernsee giant. *Lancet* **338**, 1454–5.

Schwindinger W F and Levine M A (1993) McCune-Albright Syndrome *Trends in Endocrinology and Metabolism* **4**, 238–41.

Spiegel A M (1990) Albright's hereditary osteodystrophy and defective G proteins. Editorial comment. *New England Journal of Medicine* **322**, 1461–2.

16 VENOMOUS VIPER

Introduction

Cardiovascular diseases account for about 25% of all deaths worldwide. In developed countries, about half of all deaths are due to cardiovascular disease, mainly coronary heart disease and stroke. People whose blood pressure is outside the normal range for their age face an increased likelihood of suffering a stroke or damage to the heart or kidneys. High blood pressure is a condition of unknown cause in most cases. It is a major clinical condition with no single precipitating factor. Family history of the disease indicates a higher risk but other influences such as diet (e.g. salt intake), smoking and stress are involved. It is not known which genes are involved or how they interact with environmental factors to cause the disorder. Blood pressure is a continuum and usually rises with age as resistance to flow increases in the blood vessels. In clinical terms, hypertension falls into two major categories: primary (or essential) for which there is no identifiable cause and secondary which is the result of some pre-existing disease, e.g. renal or endocrine.

As far back as the end of the last century, Tigerstedt and Bergman established that the kidney has a major role to play in the control of blood pressure. They showed that intravenous injection of an extract of rabbit kidney produced hypertension in dogs. It was later shown that this effect was due to the action of a peptide hormone secreted from the kidney. The control of blood pressure is, however, complex involving many mediators.

The present problem shows how natural products, in this case a component in snake venom, can provide important clues to physiological mechanisms such as the control of blood pressure. Naturally-occurring toxins and venoms are an important source of 'biochemical tools' and can aid in the rational design of new drugs.

Figure 16.1. Snake being 'milked' for its venom on a snake farm.

The Problem

A circulating peptide (peptide A) of M_r 1296 was isolated from human plasma and its amino acid composition determined as:

Arg,Asp,His_2,Ile,Leu,Phe,Pro,Tyr,Val

Treatment with a carboxypeptidase produced the rapid, quantitative release of Leu. Treatment with trypsin generated an octapeptide together with a dipeptide of composition Arg,Asp. Prolyl endopeptidase, a serine peptidase that cleaves on the *C*-terminal side of prolyl residues, released a heptapeptide and a tripeptide of composition His,Leu,Phe. Separate treatment of the intact peptide A with chymotrypsin produced three peptides of the following compositions:

I = His,Ile,Phe,Pro
II = His,Leu
III = Arg,Asp,Tyr,Val

When peptide I was treated with an aminopeptidase, Ile was released rapidly.

A peptidase from human lung was able to remove the *C*-terminal dipeptide from peptide A converting it into a powerful vasoactive compound (peptide B). The peptidase was inhibited by chelating agents such as EDTA and *o*-phenanthroline but not by diisopropylphosphofluoridate (DIPF) nor by thiol reagents such as iodoacetate. When lung tissue was homogenised in a buffered solution and centrifuged at 100 000 *g* for 1 hour the activity was found in the pellet but could be released in a soluble form by treatment of the pellet with the detergent Triton X-100 but not by 1 M NaCl. The solubilised enzyme could be adsorbed onto a column of an immobilised lectin, concanavalin A (Con A), and could be eluted from the column by α-methylmannoside.

In a search for selective inhibitors of the peptidase which could form the basis of antihypertensive drugs, a compound extracted from snake venom was found to be a potent, reversible inhibitor. The nature of its inhibition was examined by measuring the initial rate of reaction (v) of the enzyme at different concentrations of the venom compound as shown in Table 16.1.

Table 16.1. Inhibition of lung peptidase by snake venom compound.

	Initial rate v (nmol/min)		
Substrate (mM)	No inhibitor	0.5 μM inhibitor	1.5 μM inhibitor
0.025	0.2	0.035	0.017
0.05	0.31	0.067	0.033
0.1	0.4	0.12	0.063
0.33	0.5	0.27	0.17

Questions

1. What are the specificities of trypsin and chymotrypsin? What is the amino acid sequence of peptide A?

2. From the information given in the problem, what can you deduce about the properties of the peptidase involved in the conversion of peptide A to peptide B? (You will need to know what a chelator and a lectin are, what thiol reagents such as iodoacetate do and the significance of the Triton and NaCl experiments.)

3. By reference to a text book, make a list of peptides which are important in the regulation of blood pressure. From the information given in this problem and what you have deduced in answer to the questions so far, can you name peptides A and B and the peptidase which converts A to B?

4. From what precursor is peptide A derived and what is its role in the regulation of blood pressure?

5. What type of inhibition is exerted on the peptidase by the snake venom compound? Calculate the K_m of the substrate for the peptidase. Bearing in mind the nature of the inhibition, what general class of compound might the snake venom inhibitor belong to?

6. The venom inhibitor showed antihypertensive properties when injected but not when taken by mouth. Why might this be?

7. This inhibitor, and related snake venom compounds, have been used as a rational basis for the design of clinically-effective antihypertensive compounds. What are the requirements for such a drug?

Commentary

Trypsin and chymotrypsin are both pancreatic serine proteinases but differ in specificity. They are both inhibited by DIPF. Trypsin cleaves on the *C*-terminal side of the basic amino acids (lysine and arginine), whereas chymotrypsin cleaves after aromatic residues. By using combinations of proteolytic enzymes as well as some chemical reagents, peptides or proteins of unknown sequence can be cleaved into smaller fragments whose sequences can then be determined. The structure of the intact peptide is deduced by piecing together the various fragments generated. The problem here is a simple illustration of how this can be done although nowadays automatic protein sequencers (sequenators) can simplify these procedures. By combining the data provided, the sequence of the human peptide can be deduced as:

Asp-Arg-Val-Tyr-Ile-His-Pro-Phe-His-Leu

EDTA and *o*-phenanthroline are both metal-chelating agents suggesting that the peptidase requires a metal ion for its catalytic activity, i.e. it is a metalloproteinase. The cell fractionation data suggest that the enzyme is an integral rather than peripheral membrane protein. The evidence for this is that it is found in the particulate fraction of lung tissue (which would contain membrane fragments) and is solubilised by a detergent but not by a solution of high ionic strength (1M NaCl). Lectins are plant proteins that bind specific sugar residues with high affinity. The affinity of the peptidase for Con A implies that the peptidase is a glycoprotein. The peptidase can be eluted from the immobilised lectin by α-methylmannoside, a sugar with a high affinity for Con A, suggesting that the interaction with the lectin is specific and that the glycoprotein chains on the peptidase contain mannose residues. It is highly likely that the glycoprotein peptidase is located on the plasma membrane. The absence of any effect of DIPF or iodoacetate (which reacts with –SH groups) indicates the peptidase is not a serine peptidase and that its activity is not dependent on a critical cysteine residue.

The key peptides involved in regulation of blood pressure are angiotensin II, bradykinin, atrial natriuretic peptide and the recently discovered family of endothelin peptides. The peptide in the present problem is angiotensin I which is metabolised to the vasoactive peptide angiotensin II at the surface of lung endothelial cells through the action of angiotensin converting enzyme (ACE) which removes the *C*-terminal dipeptide His-Leu from angiotensin I. Angiotensin I is, in turn, produced from its protein precursor angiotensinogen, an α_2-globulin protein secreted from the liver. The release of angiotensin I from angiotensinogen is catalysed by the highly-specific proteinase renin which is secreted from the kidney. The only known physiological substrate for renin is angiotensinogen:

$$\text{Angiotensinogen} \xrightarrow{\text{Renin}} \text{Angiotensin I} \xrightarrow{\text{ACE}} \text{Angiotensin II}$$

Renin is a member of the class of 'aspartic proteinases' in which the carboxyl side chain of an aspartic acid plays a key role in the catalytic mechanism. Some other aspartic proteinases of physiological significance are pepsin and chymosin. An aspartic proteinase is also encoded by the human immunodeficiency virus (HIV) which is required for maturation of the viral proteins. This proteinase is therefore a therapeutic target of considerable interest in the treatment of AIDS and several such inhibitors are currently undergoing clinical trials.

The 'renin–angiotensin system' is now recognised to be one of the most important factors in the pathogenesis of hypertension. The active peptide of the cascade, angiotensin II, binds to cell-surface receptors mediating a variety of physiological effects including vasoconstriction, stimulation of aldosterone and vasopressin secretion, adrenal catecholamine release and renal sodium reabsorption. ACE also plays an additional physiological role in inactivating bradykinin, a vasodilating peptide. Because of this dual action the converting enzyme has therefore been a prime target for designing antihypertensive drugs culminating in the highly successful and widely used drug, captopril.

The nature of the enzyme inhibition by the snake venom compound can be deduced by plotting $1/v$ against $1/[S]$ at each inhibitor concentration (Lineweaver–Burk plot). The lines intercept on the vertical axis indicating competitive inhibition (Figure 16.2). The K_m of the substrate can be deduced from the intercept on the horizontal axis and is approximately 50 μM. Since the inhibition is competitive it is likely that the snake venom compound resembles the substrate in structure, i.e. is peptide in nature. This is, in fact, the case. The venom of the Brazilian pit viper, *Bothrops jararaca*, contains several small peptides that inhibit ACE. Being a peptide, the snake venom compound has limited activity by mouth since it is poorly absorbed and is hydrolysed by gut peptidases. Nevertheless the structural features that are recognised by ACE were adapted to produce the first orally active antihypertensive drug, captopril. Two important features that determined specificity were a proline residue, since peptides with a *C*-terminal proline residue produced strongest inhibition of ACE, and a thiol group which interacts strongly with the zinc ion at the active site producing an inhibitor of nanomolar potency (see Figure 16.3).

In the search for other novel classes of antihypertensive drugs there has been a lot of interest in the design of specific inhibitors of renin, although no clinically active drugs have yet come from this programme. More recently, non-peptide compounds that are

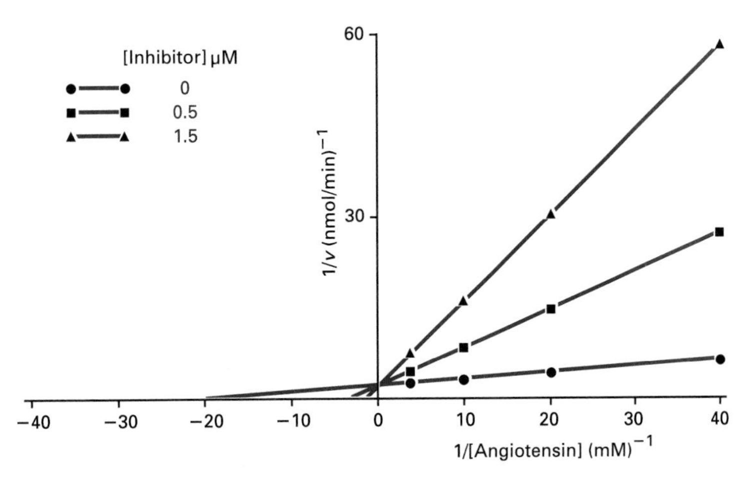

Figure 16.2. Double-reciprocal (Lineweaver–Burk) plot of the inhibition of ACE by the snake venom compound.

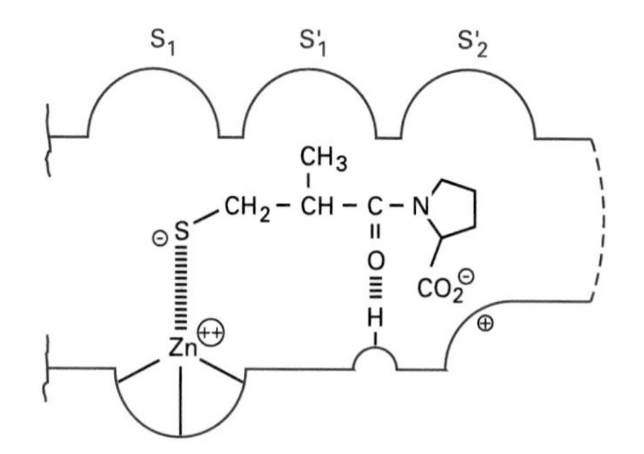

Figure 16.3. Postulated binding of the inhibitor captopril to the active site of ACE. S_1, S_1' and S_2' represent 'pockets' in the enzyme where the three C-terminal residues of a typical peptide substrate such as angiotensin I bind. The peptide bond between the amino acid residues of the substrate located in sites S_1 and S_1' is normally cleaved by the enzyme. Note that captopril is an analogue of a dipeptide (Ala-Pro). Adapted from Petrillo and Ondetti (1982) *Medical Research Reviews* **2**, 1–41.

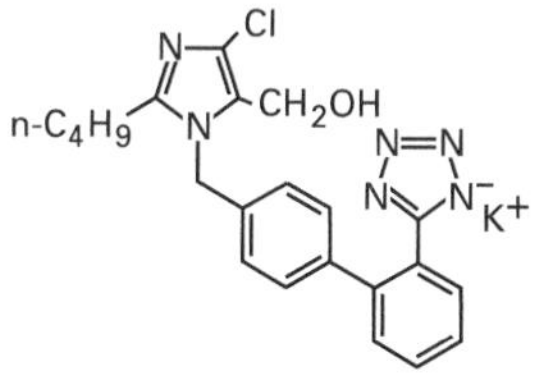

Figure 16.4. Structure of losartan, a non-peptide antagonist of the angiotensin II receptor. Adapted from Wong *et al.* (1992) *Trends in Endocrinology and Metabolism* **3**, 211–17.

potent antagonists of the angiotensin II receptor have been discovered and may have therapeutic potential (see Figure 16.4). The requirements for a clinically effective drug are oral bioavailability, chemical specificity and high affinity for the target protein, relatively few side effects and no long-term toxicity.

Much effort is currently being directed towards identifying specific genes that may predispose individuals to the development of hypertension and cardiovascular disease. For example, recent studies have shown that a deletion polymorphism in the gene encoding ACE may be a risk factor for myocardial infarction.

Further Questions

1. What other hormones are involved in the control of blood pressure and salt and water homeostasis? (See also Problem 7.)
2. What is the basis of the Edman procedure used in the automated sequencing of peptides and proteins? (See also Problem 11.)
3. How would you distinguish between competitive and non-competitive enzyme inhibition and between reversible and irreversible inhibition? (See also Problem 21.) Why is it important to measure *initial* rates of reaction in an enzyme assay?
4. Through what signal transduction mechanisms might angiotensin II, atrial natriuretic peptide and endothelins work? (See also Problem 15.)
5. Potential therapeutic targets for control of blood pressure are renin, ACE and angiotensin II receptors. What are the relative merits of each of these targets?

Connections

- Make sure you understand the difference between 'integral' and 'peripheral' membrane proteins and how you would distinguish between them.
- Use this problem to explore the sites of action of other toxins, e.g. bacterial toxins (cholera, pertussis, tetanus, botulinum), snake venoms (bungarotoxin, cobra toxin), and marine toxins (tetrodotoxin, saxitoxin). (See also Problem 13.)
- Use this problem to revise other examples of zymogens and prohormones. (See also Problem 18.)
- ACE is a zinc-containing metalloenzyme. Find other enzymes that contain zinc and list other metals that are required by enzymes. Find out what 'zinc fingers' are. (See also Problems 6 and 22.)
- ACE is also a glycoprotein. Make sure that you know what is meant by '*N*-linked' and '*O*-linked' sugars and have an overview of where and how carbohydrate is added to proteins.
- List other examples of 'serine enzymes' which are typically inhibited by DIPF. How does DIPF inhibit such enzymes? What is the 'catalytic triad' in enzymes such as trypsin and chymotrypsin?
- Review the diagnostic uses of enzymes. (See also Problems 3 and 17.)

References

Beaglehole R (1992) Cardiovascular disease in developing countries. *British Medical Journal* **305**, 1170–1.

Kurtz T W (1992) Myocardial infarction: the ACE of hearts. *Nature* **359**, 588–9. *News and views* article.

Leckie B J (1992) High blood pressure: hunting the genes. *Bioessays* **14**, 37–41.

Marx J (1991) How peptide hormones get ready for work. *Science* **252**, 779–80.

Ondetti M A, Rubin B and Cushman D W (1977) Design of specific inhibitors of angiotensin converting enzyme: new class of orally active anti-hypertensive drugs. *Science* **196**, 441–4. Original report.

Rhodes D and Klug A (1993) Zinc fingers. *Scientific American* **268**(2), 33–9.

Sharon N and Lis H (1989) Lectins as cell recognition molecules. *Science* **246**, 227–34.

Sharon N and Lis H (1993) Carbohydrates in cell recognition. *Scientific American* **268** (1), 74–81.

Timmermanns P B M W M, Wong P C, Chiu A T and Herblin W F (1991) Non-peptide angiotensin receptor antagonists. *Trends in Pharmacological Sciences* **12**, 55–62.

Vallotton M B (1987) The renin–angiotensin system. *Trends in Pharmacological Sciences* **8**, 69–74.

17 HITTING THE BOTTLE

Introduction

Alcoholic beverages have been consumed in nearly all societies throughout history and have been produced from almost any available vegetable source. Although in moderation alcohol (ethanol) may be beneficial and even provide some protection against heart disease, the medical and social effects of alcohol abuse and alcoholism are enormous. Direct medical consequences include both the acute effects of alcohol intoxication and the chronic degeneration of body functions, especially liver damage. Cirrhosis of the liver is the third most frequent cause of death among those aged 25–64 in New York City and similar urban areas. The correlation between deaths from cirrhosis of the liver and alcohol consumption in some European countries is shown in Figure 17.1. Indirect effects of alcohol abuse such as accidents and injury, especially from automobile accidents, also keep hospital casualty departments busy. Social and economic effects range from days off work and lost production to antisocial behaviour

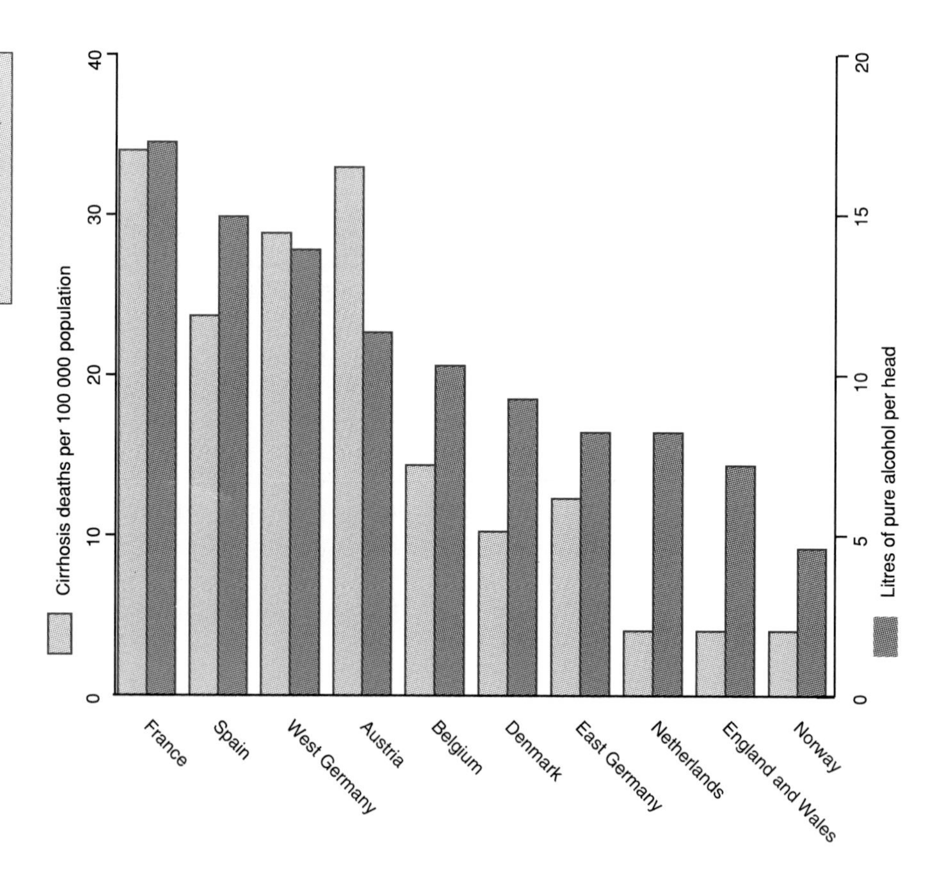

Figure 17.1. Alcohol consumption and deaths from cirrhosis of the liver in some European countries in the mid-1970s. Taken from Farrell and Strang (1992).

Table 17.1. Alcohol consumption (litres of 100% alcohol per head of population) in the European Community.

Country	1970	1990
Belgium	9.6	11.4
Denmark	7.1	9.8
France	17.3	12.6
Germany	12.1	11.9
Greece	5.3	6.0
Republic of Ireland	5.9	7.2
Italy	16.0	9.4
Netherlands	5.5	8.3
Portugal	9.8	9.6
Spain	12.1	10.7
United Kingdom	5.3	7.0

('lager louts') and violence, especially in the family. Many governments, however, do not do all they might to deal with alcohol-related problems since, as with tobacco, they derive a huge income in taxes on these products which they are reluctant to lose.

High alcohol consumption is also specifically linked with a number of cancers of the mouth, throat, oesophagus and liver. The carcinogenic effects of alcohol may be potentiated by smoking. The changes in alcohol consumption in the European Community over a twenty-year period are shown in Table 17.1. France and Italy which were the highest alcohol consuming countries in 1970 have reduced their consumption by 27% and 42%, respectively. In contrast, the UK has increased its consumption of alcohol by about 30%. The present problem is concerned with some of the metabolic and pathological consequences of alcohol abuse.

The Problem

Bill C., an unemployed labourer aged 55, had been a heavy beer drinker for years and was admitted to hospital after collapsing in the street. He was clearly unsteady on his feet, confused and with a strong smell of alcohol on his breath. His blood alcohol concentration was 78 mM. Physical examination revealed tender enlargement of the liver. He complained of general loss of appetite, fatigue, early-morning nausea and frequent gastrointestinal problems. Occasional vomiting of blood had been reported and enlarged gastro-oesophageal veins (varices) were detected. Mary P., aged 42, also presented with gastrointestinal problems and frequent diarrhoea. She had been under considerable stress at work and admitted to concerns about her alcohol consumption.

It was felt advisable to perform a set of liver function tests on both patients and the results are shown in Table 17.2. The blood calcium and magnesium levels were low and Bill's urinary urea excretion was also low. In addition, his blood clotting time was found to be impaired.

Table 17.2. Liver function tests.

Blood levels	Bill C.	Mary P.	Reference range
Total protein (g/l)	68	77	60–84
Albumin (g/l)	31	39	35–50
Total bilirubin (μM)	58	15	3–15
Alkaline phosphatase (U/l)	725	339	100–300
Alanine transaminase (U/l)	35	94	5–35
Aspartate transaminase (U/l)	42	177	10–40
γ-Glutamyl transferase (U/l)	790	463	7–45

Questions

1. Why are the particular tests shown in Table 17.2 indicative of liver function? Which patient has the more severe liver degeneration?
2. What other conditions might elevated levels of alkaline phosphatase and transaminases indicate?
3. What is bilirubin and how is it formed?
4. Why might Bill C.'s blood clotting have been deranged?
5. Some detailed histology data from a patient with alcoholic liver disease are shown in Figure 17.2 and a post-mortem liver from a case of alcoholic cirrhosis is shown in Figure 17.3. What prominent changes to the structure of the liver are evident in these figures? What are the causes and pathology of cirrhosis of the liver?

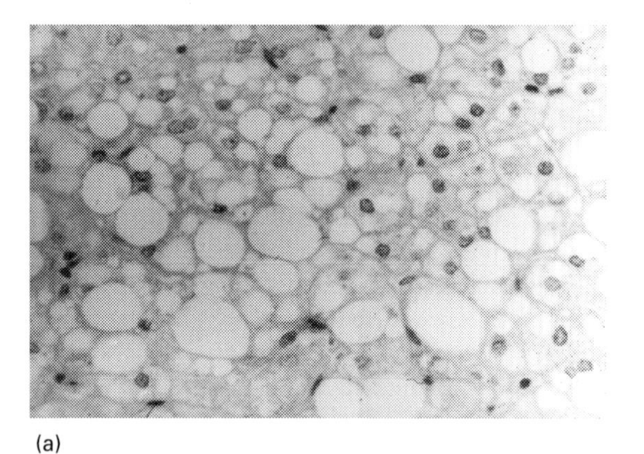

(a)

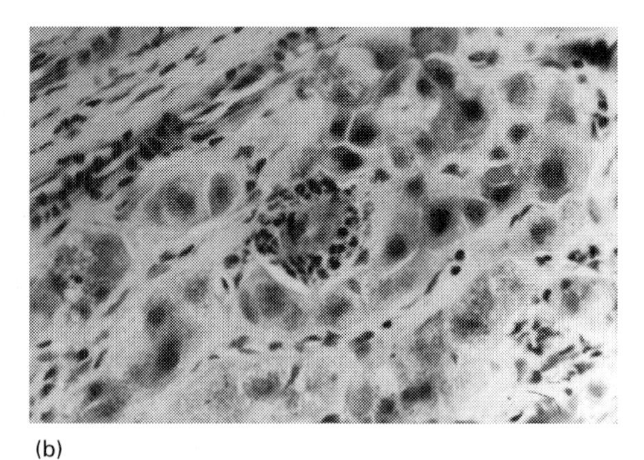

(b)

Figure 17.2. Histology of a liver biopsy sample from patient with abnormal liver function tests indicating (a), macrovesicular fatty change, and (b), the lesion of alcoholic hepatitis. Photographs kindly supplied by Dr Judy Wyatt, St James's University Hospital, Leeds.

Figure 17.3. *Post-mortem* liver from a case of alcoholic cirrhosis. Photograph kindly donated by Dr Judy Wyatt, St James's University Hospital, Leeds.

Table 17.3. Kinetic parameters of human liver aldehyde dehydrogenases towards acetaldehyde (data from Kennedy N P and Tipton K F (1990) *Essays in Biochemistry* 25, 137–95).

Subcellular compartment	K_m (μM)	V_{max} (nmol/min/mg protein)
Cytosol	126	6.9
Endoplasmic reticulum	728	9.3
Mitochondrial outer membrane	247	5.6
Mitochondrial matrix	1.4	8.1

6. Why is malnutrition often associated with chronic alcohol abuse? Why might blood calcium and magnesium levels and urinary urea excretion be low?
7. What are the main pathways of ethanol metabolism in the liver and the principal metabolic changes that occur in chronic alcohol consumption? An intermediate in the oxidation of ethanol is acetaldehyde which is oxidised to acetate by aldehyde dehydrogenase (AldDH). Several isoenzymes of AldDH occur in the liver and they are found in the cytosol, mitochondria and the endoplasmic reticulum. Their kinetic parameters are summarised in Table 17.3. Use the Michaelis–Menten equation to calculate which isoenzyme would mainly be responsible for oxidation of acetaldehyde at concentrations less than 50 μM.
8. A high dose of ethanol is sometimes used as an 'antidote' to poisoning by methanol or ethylene glycol (antifreeze). What is the biochemical basis for this treatment?
9. Why should alcohol be avoided or consumed with caution when taking other drugs?

Commentary

The term 'liver function test' is something of a misnomer since these tests do not provide a quantitative measure of liver function but merely indicate some degree of liver dysfunction. Liver function tests are useful support for initial clinical investigations but can rarely provide a precise diagnosis. Nevertheless, the tests are cheap to perform, non-invasive and particularly valuable for monitoring the progression of liver disease once a firm diagnosis has been established. The common battery of blood tests includes total protein and albumin, bilirubin, transaminases, alkaline phosphatase and γ-glutamyl transferase (GGT).

Albumin is synthesized in, and secreted from, the liver and in some chronic liver conditions may be the only detectable biochemical abnormality. However, many patients with liver disease have a normal level of serum albumin as is the case with Mary P. Increased plasma levels of the transaminases indicate leakage of the enzymes from damaged cells. The origin of plasma transaminases could be one of a number of tissues, e.g. liver, skeletal muscle, heart, kidney, and therefore the absolute levels of these enzymes lack predictive accuracy in diagnosis. The relative levels of aspartate transaminase (AST) (also known as glutamate-oxaloacetate transaminase, GOT) and alanine transaminase (ALT) (glutamate-pyruvate transaminase, GPT) can, however, be helpful in diagnosis. In the presence of deranged liver function an AST/ALT ratio of >1 is indicative of alcoholic damage.

The microstructure of the liver is shown diagrammatically in Figure 17.4. In obstruction of the bile passages and in alcoholic cirrhosis there is an increased level of secretion of alkaline phosphatase. This increase is not due to disruption of hepatocytes but to induction of synthesis of the enzyme. However, several tissue-specific isoenzymes of alkaline phosphatase exist (liver, bone, intestine, placenta). They can be distinguished by gel electrophoresis but the commonest procedure to confirm the hepatic origin of plasma alkaline phosphatase is to measure an additional enzyme which is elevated in liver disease but not, for example, in bone disease. Such an enzyme is GGT and elevated levels of GGT are found in most hepatobiliary diseases.

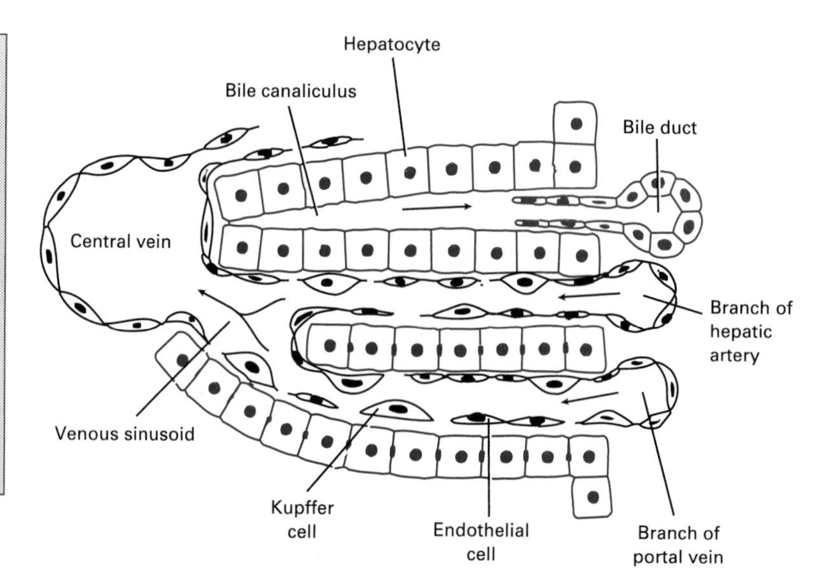

Figure 17.4. Microstructure of the liver. The liver consists of acini in which sheets of hepatocytes are permeated by sinusoids carrying blood from the portal venules and hepatic arterioles to the central vein. Bile is secreted from the hepatocytes into canaliculi, which drain into the bile ducts. Modified from Marshall (1992).

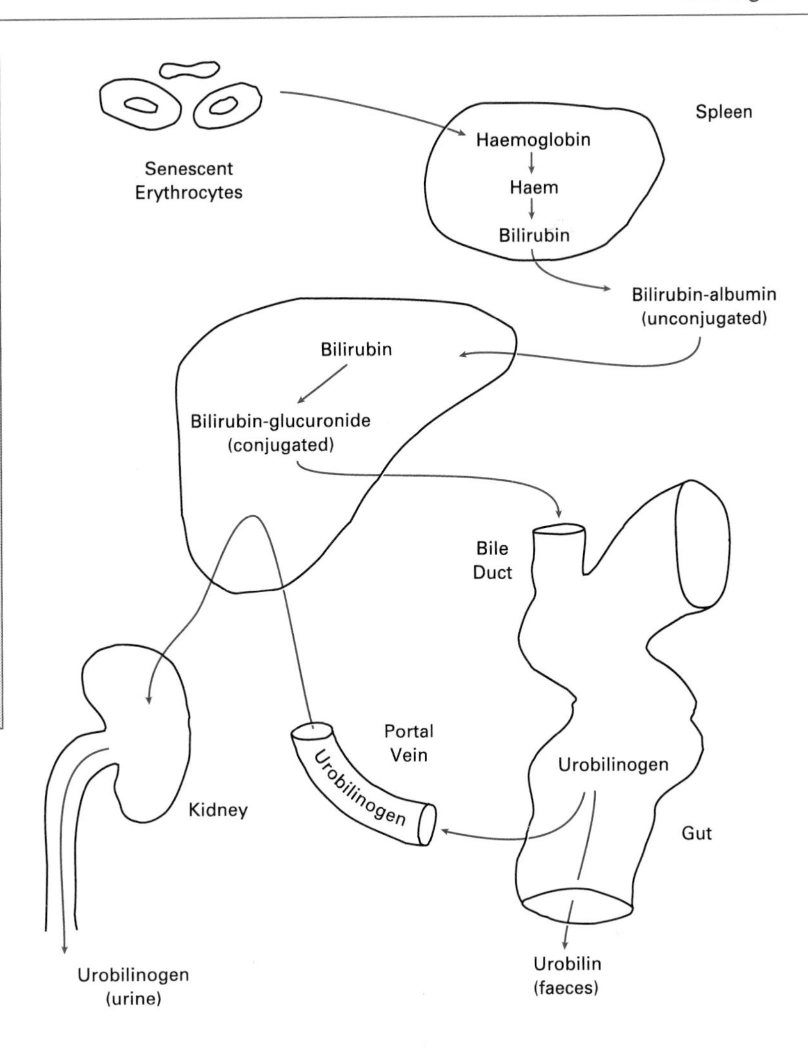

Figure 17.5. Metabolism of haem and excretion of its metabolites. Haem, the main source of which is erythrocyte haemoglobin, is metabolised first in the spleen to bilirubin. After its transport to the liver bound to albumin, the bilirubin is conjugated with two molecules of glucuronic acid and excreted into the gut via bile. In the gut, bacteria further metabolise the bilirubin. Some urobilinogen is reabsorbed and eventually excreted in the urine.

GGT is often used as a marker for alcohol consumption since its synthesis is induced by ethanol. When used in such a context, the assay of GGT is sensitive but not very specific since the enzyme is also induced by a wide variety of drugs including some anti-epileptics and oral contraceptives. The levels of glutathione *S*-transferase may be a more sensitive and specific enzyme indicator of liver damage but the assay is not routinely performed.

Hyperbilirubinaemia is not always a feature of chronic liver disease and can also be seen in some other disease states. The serum concentration of bilirubin is determined by the relative rates of its synthesis and clearance. As shown in Figure 17.5, bilirubin mainly comes from degradation of the tetrapyrrole ring of haem in senescent red blood cells and can occur in two forms: unconjugated and conjugated. Unconjugated bilirubin, which is a lipophilic molecule, is transported in the blood bound to albumin. It is then taken up by the liver hepatocytes and undergoes conjugation with glucuronic acid to form a hydrophilic diglucuronide conjugate which is readily excreted into bile. In a normal person nearly all the plasma bilirubin is in the unconjugated state. Clinical assays of bilirubin are usually expressed as 'direct' and 'indirect' which are approximate measures of conjugated and unconjugated bilirubin respectively. Bilirubin can also occur in a form covalently linked to albumin and this may be a major component in some forms of hepatic disease. Although the serum bilirubin concentration can be a very specific test for liver disease, it is not particularly sensitive since the capacity of the liver to conjugate bilirubin exceeds its production from haem degradation by several-fold.

In the case of Bill C., the abnormal clotting, the high bilirubin and alkaline phosphatase levels, but with transaminase levels at the upper end of the normal range, suggest serious derangement of liver function, probably advanced cirrhosis. This is also supported by the oesophageal varices and vomiting of blood which arise from portal hypertension due to cirrhosis. Fibrosis of the liver has led to distortion of the intrahepatic biliary tree, hence the elevated bilirubin and alkaline phosphatase levels. The substantial loss of normal, functioning hepatocytes means that the plasma levels of the transaminases are not much above normal. Derangement of clotting occurs due to decreased synthesis of clotting factors by the liver.

Mary P., who has moderate alcoholic liver disease, shows a different pattern. In her case, GGT is a sensitive and early indicator of liver dysfunction and the transaminases are also abnormally raised, although alkaline phosphatase and bilirubin levels are only slightly above normal.

Even relatively low levels of alcohol can produce ultrastructural changes in hepatocytes, including the accumulation of fatty droplets in the liver and the proliferation of the smooth endoplasmic reticulum. Enlargement of mitochondria is also often seen. Chronic alcohol abuse inevitably leads to hepatic steatosis ('fatty liver') due to the accumulation of triglycerides. Patients with steatosis may be without specific symptoms apart from some loss of appetite accompanied by early morning nausea. In some susceptible individuals steatosis may progress to the development of alcoholic hepatitis.

Figure 17.2a shows the macrovesicular changes which would be commonly seen in a liver biopsy from a patient with abnormal liver function tests who drank alcohol to excess. However, the changes seen are rather non-specific and they may be seen in other circumstances such as obesity, diabetes or treatment with drugs such as steroids. Figure 17.2b shows a lesion typical of alcoholic hepatitis – a hepatocyte containing fibrillar structures known as Mallory's Bodies (Mallory's hyaline) surrounded by neutrophils which are seen at the centre of the field. Adjacent hepatocytes show pericellular fibrosis. This is the less frequent but much more specific feature of alcoholic liver disease which may well progress to cirrhosis. In cirrhosis, which occurs in only 10–15% of alcoholics, the lobular structure of the liver is destroyed and there is progressive infiltration of fibrous tissue between nodules of functional hepatocytes (Figure 17.3). Bile duct involvement may occur. The presence of oesophageal varices is a late feature of cirrhosis and there may also be signs of gonadal dysfunction (alcoholic feminisation). Alcohol abuse is a major but not the only cause of cirrhosis of the liver which is the end stage of any chronic liver disease such as viral or autoimmune hepatitis as well as some cases of inborn errors of metabolism.

Chronic alcoholics usually do not consume a balanced diet, taking in most of their calories in the form of ethanol. They are therefore at considerable risk of nutritional,

Figure 17.6. Alcoholic liver disease. From Kaplan and Pesce (1989) *Clinical Chemistry: Theory, Analysis and Correlation.* C V Mosby, St. Louis..

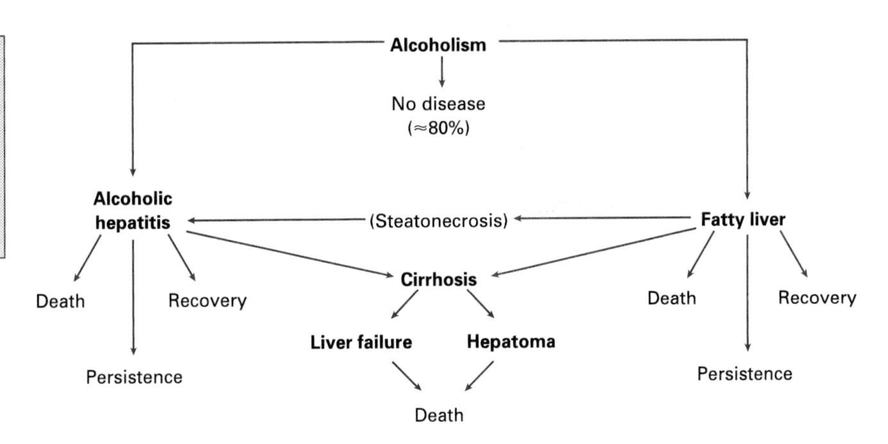

especially vitamin and mineral, deficiencies. Alcohol consumption also produces damage to the gastrointestinal tract resulting in impairment of the absorption of certain nutrients. Vitamin K deficiency may be contributing to Bill C.'s poor blood clotting. The absorption of the fat-soluble vitamin K from the gut is compromised when bile production and excretion are impaired in liver disease. Neurological symptoms associated with deficiencies in thiamine or pyridoxine and haematological problems because of folate or pyridoxine deficiencies are common in alcoholics. Thiamine deficiency can lead to the Wernicke–Korsakoff syndrome whose symptoms include ataxia and mental confusion. Severe cases of thiamine deficiency produce the condition known as beri-beri. The hypomagnesaemia and hypocalcaemia seen in the case of Bill C. are probably a consequence of poor diet and impaired uptake of the ions by the gut. This can, in turn, lead to decreased circulating levels of parathyroid hormone (PTH) since magnesium is required for the synthesis and action of PTH. The diminished action of PTH on the renal tubules would result in low urinary cAMP levels. Chronic alcohol abuse may also derange bone metabolism and cause osteoporosis. The reduced urinary urea excretion seen in Bill C. is also a result of poor diet and impaired protein metabolism. The urea cycle, which is located in the liver, may also be affected. At one time alcohol-induced disease was attributed exclusively to the associated malnutrition. More recent studies have shown that fatty liver and cirrhosis can be produced by alcohol in the absence of dietary deficiencies and even in the presence of enriched diets. The development of liver damage due to alcohol abuse is shown in Figure 17.6.

The primary site of metabolism of ethanol is the liver and the initial event is the NAD^+-dependent oxidation of the alcohol by cytosolic alcohol dehydrogenase (ADH):

$$CH_3CH_2OH + NAD^+ \rightleftharpoons CH_3CHO + NADH + H^+$$

Multiple ADH isoenzymes exist, there being five different genes encoding the enzyme in humans. The frequency of expression of the ADH alleles differs among different ethnic groupings. Although ADH is the principal pathway of ethanol oxidation in liver, additional pathways can operate at high ethanol concentrations. These include oxidation by a cytochrome P_{450}-dependent pathway in the endoplasmic reticulum (the microsomal ethanol-oxidising system) and by catalase. The product in all cases is acetaldehyde which is further oxidised to acetate by NAD^+-dependent aldehyde dehydrogenase (AldDH). The enzyme located in the mitochondrial matrix would account for most of the acetaldehyde oxidation at concentrations of less than 50 μM. This can be calculated by substituting the appropriate kinetic parameters listed in Table 17.3 into the Michaelis–Menten equation:

$$\text{Initial rate } (v) = \frac{V_{\max}[S]}{K_m + [S]}$$

where K_m = Michaelis constant, [S] = substrate concentration, and $V_{\max}$ = maximal reaction velocity.

The metabolism of ethanol in the liver results in a marked increase in the $NADH:NAD^+$ ratio which, in turn, produces a decrease in the pyruvate:lactate ratio. The increased concentration of lactate can lead to hyperlactic acidaemia which reduces the capacity of the kidney to excrete uric acid which, in turn, may lead to secondary hyperuricaemia. In susceptible individuals this may aggravate or precipitate attacks of gout. The increased $NADH:NAD^+$ ratio can also cause a decrease in gluconeogenesis. The storage of glycogen is diminished in alcoholics through poor dietary intake and liver disease. Hypoglycaemia may occur in association with alcohol abuse and a relatively normal liver. Alcohol-induced fasting hypoglycaemia occurs in chronically-malnourished alcoholics or when moderate to large amounts of alcohol are consumed after a 6–36 hour fast. Particularly vulnerable are children who find dad's whisky bottle! The change to a more reduced state within the hepatocyte results in an inhibition of fatty acid catabolism and therefore an increased accumulation of fats in the form of triglycerides, hence the development of fatty liver. In addition, metabolism of ethanol produces an excess of acetate units which are diverted into fatty acid synthesis. Remember that the liver is the major site of fatty acid metabolism and the synthesis of the various lipoprotein complexes found in the circulation. However, some of the acute and chronic effects of alcohol are directly due to the acetaldehyde formed rather than the change in the $NADH:NAD^+$ ratio.

Ethylene glycol and methanol are also oxidised by ADH although they are much poorer substrates than ethanol itself which is preferentially oxidised. It is the oxidation products of ethylene glycol (oxalate) and methanol (formaldehyde) which are toxic. Consumption of a large dose of ethanol will inhibit oxidation of the ethylene glycol or methanol, allowing the potentially toxic compounds to be excreted unchanged. In this case ethanol is acting as a 'competitive substrate'. Hydroxylation of drugs which are metabolised by the microsomal P_{450} system in the liver is inhibited by ethanol thereby prolonging the time they remain in the body as well as their circulating levels. Thus alcohol consumption should generally be avoided when taking such drugs. Alcohol can directly potentiate the depressant action of some drugs, e.g. barbiturates or benzodiazepines (Valium, Librium) which can be hazardous. All three classes of drugs mediate some or all of their actions through the receptor for the inhibitory neurotransmitter γ-aminobutyrate (GABA). A particular subtype of GABA receptor in the cerebellum has been implicated in some of the behavioural effects and motor impairment seen in alcohol intoxication.

Further Questions

1. What are the major metabolic functions of the liver?
2. From what other proteins, apart from haemoglobin, is bilirubin derived?
3. In the formation of bilirubin diglucuronide, what is the donor of glucuronic acid and the enzyme involved?
4. What is the biochemical role of vitamin K?
5. What is the biochemical role of thiamine?
6. The compound disulfiram (Antabuse), which is an inhibitor of aldehyde dehydrogenase, has been used to prevent alcohol consumption by chronic alcoholics. How do you think that this compound would deter alcoholics from drinking alcoholic beverages?
7. Orientals often exhibit extreme sensitivity to ethanol resulting in palpitations, nausea and the 'flushing reaction', in which facial flushing occurs after a relatively small intake of alcohol. Why might this be?
8. Subcellular fractionation of liver tissue has revealed the existence of distinct AldDH isoenzymes in different subcellular compartments. What enzymes or other markers are characteristic of the major compartments within the cell, e.g. nucleus, mitochondria, endoplasmic reticulum, lysosomes, Golgi complex, plasma membrane, cytosol?
9. U.S. legislation now requires a health warning on liquor bottles to pregnant women. Why should pregnant women minimise their alcohol consumption? Find out what you can about 'foetal alcohol syndrome' which is probably the leading cause of mental retardation and neurological abnormality in the Western world today.

Connections

- Make sure you understand the various pathways by which drugs are detoxified, especially the cytochrome P_{450} system. Sometimes in its attempts to detoxify a compound the body inadvertently generates a much more toxic product. List some examples. (See also Problem 14.)

- Recall the process of haem degradation and possible causes of jaundice. (See also Problem 11.)
- Make sure you understand the mechanisms that regulate mineral homeostasis and the various hormones involved in this process. (See also Problem 15.)
- Make a list of enzymes that can be used in clinical diagnosis and make sure you understand how different tissue-specific isoenzymes can be distinguished. (See also Problem 3.)
- Review the control of lipid metabolism, especially triglyceride synthesis and transport. (See also Problem 20.)
- Ensure you understand how different types of enzyme inhibition can be distinguished. (See also Problem 16.)
- Use this problem to remind yourself of the metabolic origin of uric acid and its link with gout. Make sure you understand the biochemical basis of treatment of gout with allopurinol.

References

Abel L L and Sokol R J (1986) Fetal alcohol syndrome is now the leading cause of mental retardation. *Lancet* **2**, 1222.

Farrant M and Cull-Candy S (1993) GABA receptors, granule cells and genes. *Nature* **361**, 302–3. *News and views* article dealing with GABA receptors, benzodiazepines and alcohol.

Farrell M and Strang J (1992) Alcohol and drugs. *British Medical Journal* **304**, 489–91.

Holden C (1992) Fetal alcohol syndrome. *Science* **258**, 739.

Johnson P J (1989) Role of the standard 'liver function tests' in current clinical practice. *Annals of Clinical Biochemistry* **26**, 463–71.

Kennedy N P and Tipton K F (1990) Ethanol metabolism and alcoholic liver disease. *Essays in Biochemistry* **25**, 137–95.

Leiber C S (1988) Biochemical and molecular basis of alcohol-induced injury to liver and other tissues. *New England Journal of Medicine* **319**, 1639–50.

Marshall W J (1992) The liver, Chapter 6 *and* Clinical enzymology, Chapter 17. In *Clinical Chemistry* (2nd edn), Gower Medical Publishing, London.

18 LACK OF CLEAVAGE

Introduction

Diabetes mellitus affects 1–2% of people in many populations and is closer to 5% in the USA; its incidence is rising. It is probably the third biggest killer after heart disease and cancer. The account of the disease by Aretaeus of Cappadocia in the 2nd Century AD is highly descriptive:

> Diabetes is a remarkable disorder. It consists of a moist and cold wasting of the flesh and limbs into urine. The disease is chronic in character and is slowly engendered, though the patient does not survive long when it is completely established, for the marasmus produced is rapid, and death speedy.

The polyuria that is characteristic of diabetes led to the origin of the name, from the Greek word meaning 'a siphon'. The Latin appendage 'mellitus' (meaning honeyed) was added by physicians in the 17th Century and relates to the sweet taste of the urine and serum of diabetics. These days, since the isolation of insulin in 1921 by Frederick Banting and Charles Best, diabetes can be controlled. However, the life expectancy of sufferers is probably reduced by one-third, on average, with long-term complications such as renal and cardiovascular disease developing. Diabetes is also the major cause of blindness in many countries. Not all diabetics require insulin to survive. Insulin-dependent (or type I) diabetics represent less than 20% of the total. The remainder (non-insulin-dependent or type II diabetics) are more resistant to the effects of insulin than non-diabetics and can generally be treated by dietary control together with the use of oral hypoglycaemic agents such as sulphonylureas.

As shown in Figure 18.1, insulin is a polypeptide hormone consisting of two disulphide-linked chains, the A chain of 21 amino acids and the B chain of 30 amino acids. Insulin was the first protein to be completely sequenced and the first to be synthesized chemically. It is synthesized in the β-cells of the pancreatic islets as a single chain precursor polypeptide, proinsulin, which is proteolytically cleaved at pairs of basic residues to generate the active hormone by one or more serine proteinases resembling the bacterial enzyme subtilisin. Insulin exerts its actions through a cell-surface receptor which is one of the family of protein tyrosine kinase receptors showing similarity with the epidermal growth factor receptor and the src family of oncogene products. The receptor, like insulin itself, is composed of two types of subunits, α and β, initially synthesized as a single $\alpha\beta$ precursor polypeptide and assembled as a functionally active tetramer, $\alpha_2\beta_2$. The α subunit (M_r 135 000) is the insulin-binding subunit whereas the transmembrane β subunit (M_r 90 000) possesses the tyrosine kinase activity which mediates the intracellular signalling.

To understand the metabolic disturbances in diabetes requires an understanding of nutrient homeostasis and the anabolic role of insulin. Insulin promotes the rapid uptake of nutrients such as glucose and some amino acids into tissues and its absence has often been likened to 'starvation in the midst of plenty'. The present problem focuses on the molecular basis for one severe and inherited form of insulin resistance.

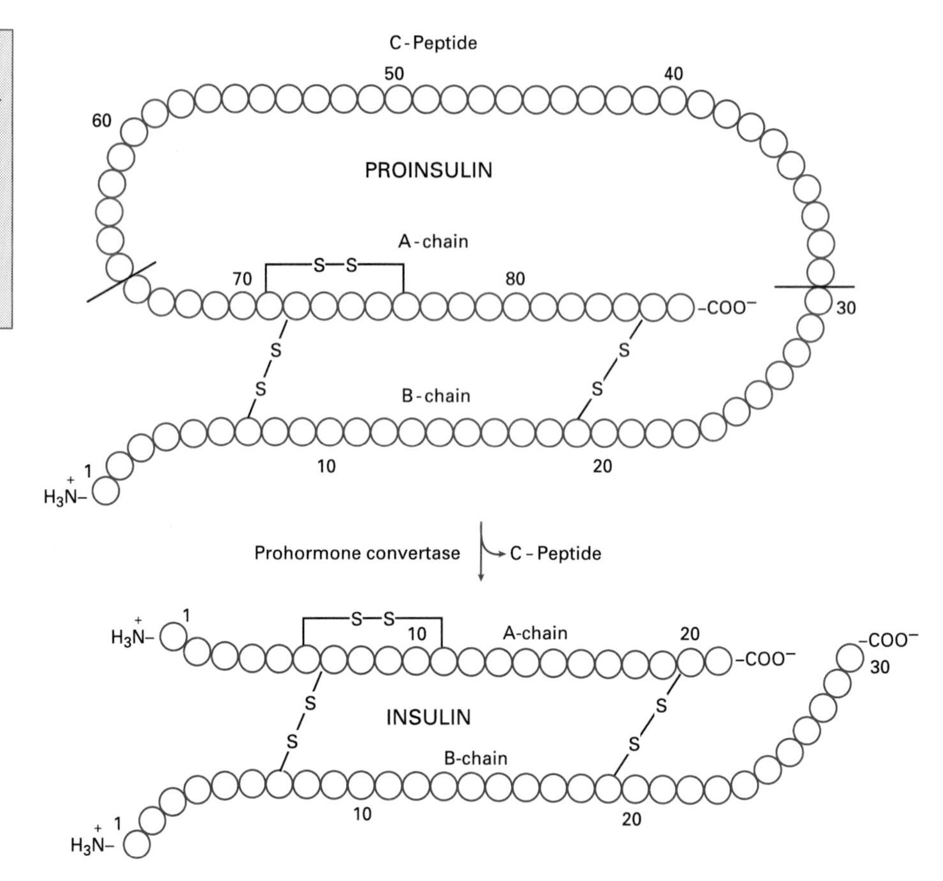

Figure 18.1. Diagrammatic representation of human proinsulin and insulin. Adapted from Devlin (1992) *Textbook of Biochemistry with Clinical Correlations* (3rd edn). Wiley-Liss, New York.

The Problem

A 23-year-old Chinese woman (Suzie Q.) had been diagnosed as diabetic at the age of seven. She had severe hyperinsulinaemia and exhibited many of the typical features of major insulin resistance including primary amenorrhoea, acanthosis nigricans (a hyperpigmented and hyperkeratotic skin rash), hirsutism and virilisation. Table 18.1 indicates the severity of some of the metabolic changes. Suzie also exhibited additional symptoms not normally associated with insulin resistance including mental retardation, short stature and dental dysplasia.

In order to investigate the molecular mechanism underlying Suzie's insulin resistance, cultured fibroblasts from Suzie were compared with those from normal subjects in their ability to bind [^{125}I]insulin (Figure 18.2a). In addition, the ability of insulin to stimulate the uptake of the non-metabolisable sugar, 2-deoxyglucose (2-DOG), into the cultured fibroblasts was examined (Figure 18.2b).

To characterise further Suzie's insulin receptors, a sample of her fibroblasts and those of a control subject were separately treated with [^{125}I]insulin together with a cross-linking reagent to couple bound insulin covalently to its receptors. The cells were then solubilised and the insulin receptors immunoprecipitated with an antibody directed against the human insulin receptor. The precipitate was then subjected to polyacrylamide gel electrophoresis in the presence of the anionic detergent sodium

Table 18.1. Changes of blood insulin and glucose concentrations in Suzie Q.

	Suzie Q.	Normal range
Fasting blood insulin (pM)	2120	20–100
Blood insulin after oral glucose loading (pM)	7120	144–502
Fasting blood glucose (mM)	7–12	3.5–5.5

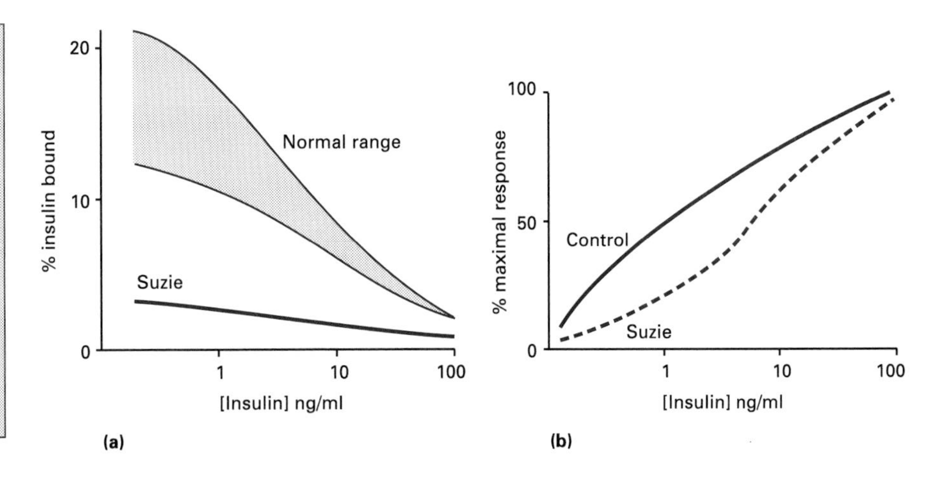

Figure 18.2. (a) Insulin binding to fibroblasts from Suzie Q. and control subjects. The assay measures the ability of increasing concentrations of unlabelled insulin to displace a limiting amount of [^{125}I]insulin. (b) Effects of insulin on the uptake of 2-DOG into fibroblasts from Suzie Q. and a control.

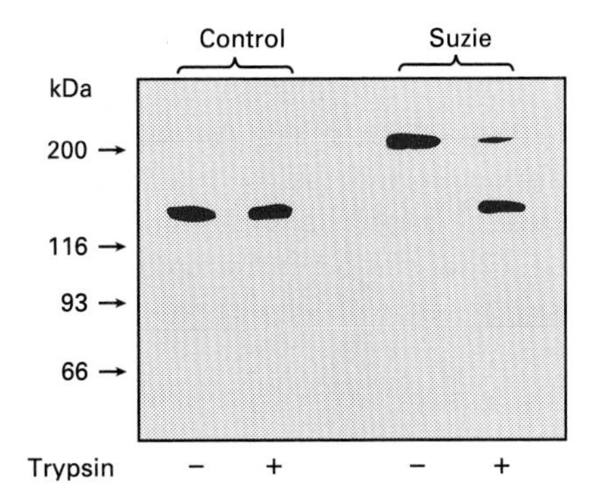

Figure 18.3. Autoradiograph of SDS-PAGE gel showing radiolabelled polypeptides.

dodecyl sulphate (SDS–PAGE); a thiol compound (dithiothreitol) was also included. The same experiment was also performed after brief treatment of the cells with a low concentration of trypsin. An autoradiograph of the SDS-PAGE gel is shown in Figure 18.3. Next, the effect of trypsin on the binding of [^{125}I]insulin to Suzie's fibroblasts was compared with those of a control. Marked differences were seen as shown in Figure 18.4.

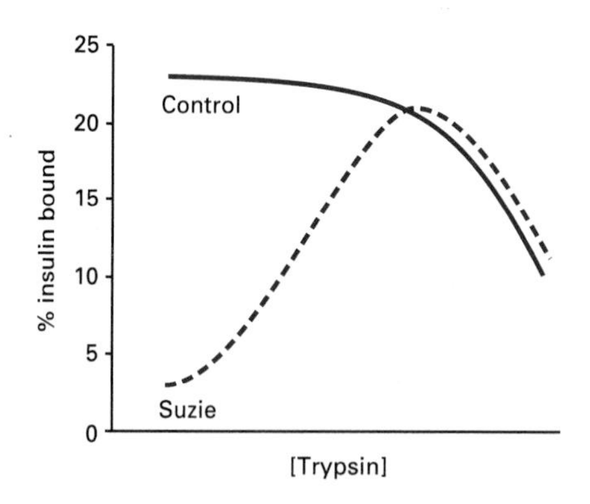

Figure 18.4. Effect of trypsin on insulin binding to Suzie Q.'s fibroblasts compared with a control.

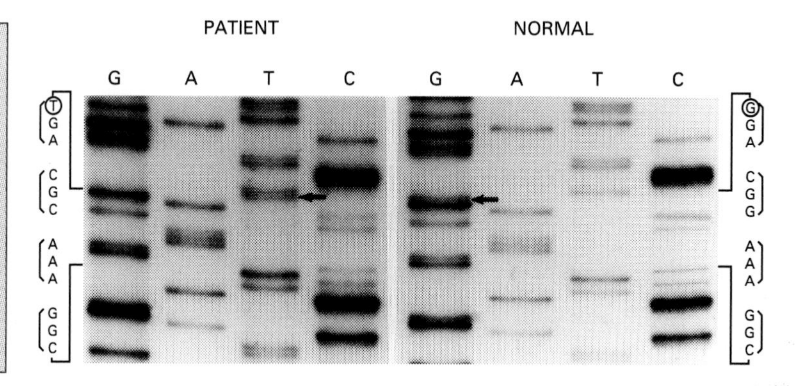

Figure 18.5. Comparison of the sequence of a region of the insulin receptor gene from Suzie Q. compared with a normal allele. ← indicates the one-base mutation. Adapted from Yoshimasa *et al.* (1988) *Science* **240**, 784–90.

In contrast to the data on Suzie's insulin receptor protein, no differences were found in the cellular levels and size of her insulin receptor mRNA transcripts compared with a normal subject following a Northern blot analysis. In a more detailed study, exon 12 of the insulin receptor gene (275 bp) was isolated and sequenced. The resultant sequence of a region of this exon is compared between Suzie and a control in Figure 18.5.

Questions

1. What are the distinctions between type I (insulin-dependent) and type II (non-insulin-dependent) diabetes mellitus?

2. How might diabetes be diagnosed?

3. What information can you derive from the data in Figures 18.2a and 18.2b about the functioning of Suzie's insulin receptor?

4. What are the principles of SDS–PAGE and what is the significance of the added dithiothreitol? What further information can be derived from an examination of the effects of trypsin on Suzie's insulin receptor compared with the control shown in the SDS gel in Figure 18.3?

5. How do you explain the initial large increase in insulin binding seen on trypsin treatment of Suzie's fibroblasts and why does binding decline rapidly as the trypsin concentration increases further (Figure 18.4)?

6. From the DNA sequencing gel shown in Figure 18.5, compare the derived amino acid sequences of this region of the insulin receptor and deduce the amino acid change in Suzie's insulin receptor. The Genetic Code is in Appendix 6. How might this change explain the differences in response between Suzie's receptor and that of the control subject?

7. Can you suggest possible explanations for some of the symptoms with which Suzie presents?

Commentary

Diabetes mellitus is not a single disease but the collective name for a group of disorders in which persistent hyperglycaemia is seen, both in the fasting state and after meals. The two major classifications of diabetes are into insulin-dependent (type I; juvenile onset) and non-insulin-dependent (type II; maturity onset). Some women also experience abnormal glucose tolerance during pregnancy, so-called gestational diabetes.

As its name suggests, juvenile-onset diabetes occurs most frequently in children and young adults with an abrupt onset which may be life-threatening and which requires the immediate administration of insulin. This sudden collapse of glucose homeostasis, however, is the end-point of an insidious process of destruction of pancreatic β-cells which may have taken place over several years. Type II diabetes is generally less severe and is usually seen after the age of 30. Insulin production is not often a problem in this class of diabetics but, rather, it is a failure to respond adequately to the circulating hormone. Table 18.2 summarises some of the major distinctions between the two types of diabetes.

Most diabetics will show fasting blood glucose concentrations of more than 7 mM. It may be difficult to diagnose type II diabetes without resort to a glucose-tolerance test. After an overnight fast, patients are given glucose by mouth (typically 75 g) and the blood glucose concentration is then measured at intervals over the next 2 hours. Typical responses are as shown in Figure 18.6.

Table 18.2. Characteristics of insulin-dependent and non-insulin-dependent diabetes.

	Insulin-dependent	**Non-insulin-dependent**
Alternative terminology	type I; juvenile onset	type II; maturity onset
Percentage of diabetics	<20%	>80%
Family history of diabetes	uncommon	common
Age at onset	usually <30 years	usually >30 years
Appearance of symptoms	rapid	slow
Obesity at onset	rare	common
Ketoacidosis	frequent	rare
Blood insulin	low	low, normal or increased
Islet β-cells	markedly decreased	normal, or slight decrease
Inflammatory islet cells	present initially	absent
Antibodies to islet cells	present	absent
HLA association	yes	no

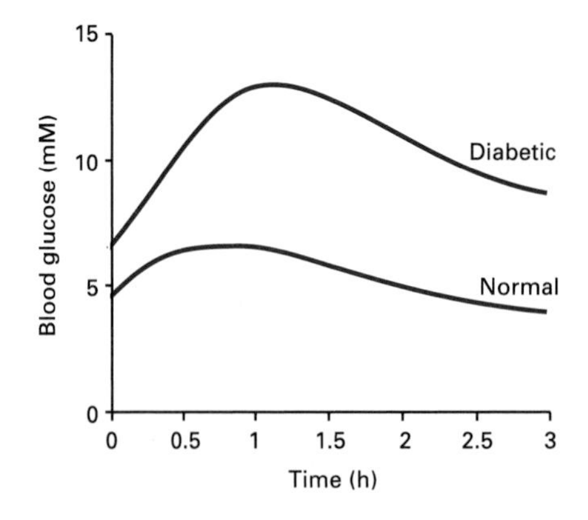

Figure 18.6. Typical responses of a normal subject and a mild diabetic to a glucose tolerance test.

Although fat cells would be an ideal choice for study of insulin actions, obtaining a patient sample would require an invasive procedure. Skin fibroblasts are therefore often used as a convenient and accessible source of cells from a patient to investigate abnormalities of insulin receptors although they have limitations and are not 'typical' insulin-responsive cells. In order to 'immortalise' these cells for extended studies in culture they would be transformed with, for example, Epstein–Barr virus, the virus implicated in 'glandular fever' (infectious mononucleosis) and nasopharyngeal carcinoma. The data of Figure 18.2a reveal that the patient's fibroblasts bind insulin far less efficiently than those of normal subjects. Among the rapid actions of insulin is the stimulation of the uptake of glucose and some amino acids into insulin-responsive cells. In the experiment shown in Figure 18.2b this effect is conveniently measured by using the non-metabolisable sugar 2-deoxyglucose. The decreased insulin binding demonstrated in Figure 18.2a is reflected in the dose–response curve for insulin stimulation of sugar uptake which is shifted markedly to higher insulin concentration, i.e. the patient exhibits *insulin resistance.*

Polyacrylamide gel electrophoresis (PAGE) in the presence of the anionic detergent SDS (SDS–PAGE) allows proteins to be separated on the basis of their molecular size. The SDS denatures the proteins and binds uniformly to the unfolded polypeptide chains producing a constant negative charge per unit length. The proteins can then be separated by electrophoresis through a cross-linked slab of polyacrylamide gel with their mobility dependent purely on the basis of their molecular mass. The migration of all the proteins is towards the anode with small proteins moving furthest since they are able to penetrate the pores in the gel more readily. SDS–PAGE produces a rapid and sensitive method of separating proteins with a high degree of resolution, as well as providing an estimate of their apparent M_r. The inclusion of dithiothreitol causes the reduction of protein disulphide bonds hence separating polypeptide chains that are covalently linked in native proteins. The SDS–polyacrylamide gel shown in Figure 18.3 reveals that the insulin-binding (α) subunit of the insulin receptor has a M_r of approximately 135 kDa and that this is unchanged in the presence of low concentrations of trypsin. In marked contrast, Suzie's insulin receptor migrates with a M_r of greater than 200 kDa but this can be converted to the 'normal' size of 135 kDa for the insulin-binding subunit by mild trypsin treatment. Note that Figure 18.3 shows an autoradiograph which reveals the location of ^{125}I. Thus, only the insulin-binding portion of the receptor (α subunit), to which [^{125}I]iodoinsulin has been covalently coupled, is revealed.

Although low levels of trypsin have no effect on the insulin-binding capacity of the insulin receptor from the normal subject, increasing concentrations result in a rapid loss in binding capacity (Figure 18.4), presumably due to proteolysis and general degradation of the receptor. On trypsin treatment of Suzie's insulin receptor, however, there is initially a marked increase in insulin-binding capacity which reaches control levels before declining as trypsin levels are increased further. Taken together, the data so far would suggest that Suzie's insulin receptor exists as the less active proreceptor and can be processed to a form resembling the normal receptor by mild trypsin treatment thereby restoring its full insulin-binding capacity.

The sequencing gel shown in Figure 18.5 provides a further clue to the changes in the properties of Suzie's insulin receptor. The point mutation of a guanine to a thymine in the coding strand of the gene changes the amino acid sequence in this region of the receptor from:

-Arg-Lys-Arg-Arg- → -Arg-Lys-Arg-Ser-

The arginine residue that is changed in Suzie's receptor is presumably an important recognition site for the processing enzyme that normally cleaves the proreceptor into its active α and β subunits. Treatment of Suzie's receptor with trypsin mimics the normal activation process by cleaving the receptor precursor on the carboxyl side of one or more of the adjacent basic residues.

In conclusion, therefore, Suzie's insulin-resistant diabetes is due to a point mutation in her insulin receptor gene that prevents the normal processing of the insulin proreceptor protein. However, the proreceptor retains some residual but weak biological activity and is therefore not an inactive zymogen. A similar multibasic processing site also occurs in the envelope protein precursor of the human immunodeficiency viruses HIV-1 and HIV-2 and in the influenza virus haemagglutinin protein. Dibasic processing sites are very common among precursors of peptide hormones (e.g. insulin) and neuropeptides (e.g. enkephalins).

Extreme insulin resistance is often associated with *acanthosis nigricans*, probably caused by direct toxic effects of the increased insulin levels on the skin. Premenopausal women with extreme insulin resistance often show features of virilisation due to insulin mediating increased ovarian production of testosterone. Insulin resistance can also result in growth retardation *in utero* seen in its most extreme form as 'leprechaunism'. Other growth and developmental abnormalities may be due to the high insulin concentrations acting on other, structurally-related receptors, e.g. the insulin-like growth factor (somatomedin) receptors. Finally, note that this particular case of diabetes is extremely rare and, strictly speaking, cannot be classified as type I or type II.

Further Questions

1. Insulin is the product of the β-cells of the pancreatic islets. What are the products of the other major classes of islet cells?

2. What factors regulate insulin secretion? How are drugs such as the sulphonylureas believed to potentiate insulin secretion?

3. What are the major catabolic hormones that oppose the anabolic actions of insulin?

4. Ketoacidosis is common in type I diabetes. What are the 'ketone bodies' and how do they arise? (See also Problem 10.)

5. How far can the underlying biochemistry explain some of the long-term complications of diabetes, e.g. peripheral neuropathy and cataract formation? (You will need to consider the role of aldose reductase.)

6. Type I diabetes is probably an autoimmune mediated disorder. What other autoimmune conditions do you know of?
7. Through what signalling mechanism(s) does insulin operate?

Connections

- Review the principal metabolic actions of insulin.
- Type I diabetes shows an association with the human leukocyte associated (HLA) antigens. Find out what you can about these and their relationship to the major histocompatibility complex (MHC) cluster of genes.
- Insulin stimulates glucose transport by increasing the V_{max} for the transport process but not the K_m. Explore the mechanism involved. (See also Problem 21.)
- What method would you use to measure blood insulin levels? (See also Problem 12.)
- Insulin, when bound to its cell-surface receptor, can be internalised by the process of receptor-mediated endocytosis. Make sure you know the mechanism involved and the role of clathrin-coated pits. (See also Problems 9 and 20.)
- Make sure you understand in what ways the metabolic changes seen in diabetes resemble those seen in starvation. (See also Problem 2.)
- What other proteins do you know of that are synthesized initially in an inactive zymogen form? Where in the cell does this processing take place? (See also Problem 16.)
- Use this problem as a basis for reviewing the actions of other important hormones. (See also Problems 7, 12, 15 and 22.)
- Explore the role of protein-tyrosine kinases in cell signalling and control of cell growth.

References

Atkinson M A and Maclaren N K (1990) What causes diabetes? *Scientific American* **263** (1), 42–9.

Diamond J M (1992) Diabetes running wild. *Nature* **357**, 362–3. *News and views* article.

Espinal J (1989) *Understanding Insulin Action: Principles and Molecular Mechanisms.* Ellis Horwood, Chichester.

Marx J (1991) How peptide hormones get ready for work. *Science* **252**, 779–80.

Orci L, Vassalli J-D and Perrelet A (1988) The insulin factory. *Scientific American* **259** (3), 50–61.

Steiner D F, Smeekens S P, Ohagi S and Chan S J (1992) The new enzymology of precursor processing endoproteases. *Journal of Biological Chemistry* **267**, 23 435–8.

Taylor S I, Cama A, Kadowaki H, Kadowaki T and Accili D (1990) Mutations of the human insulin receptor gene. *Trends in Endocrinology and Metabolism* **1**, 134–9. For a more detailed review, see Taylor S I *et al.* (1992) *Endocrine Reviews* **13**, 566–95.

19 AMINO ACID OVERLOAD

Introduction

Amino acids are the building blocks for proteins, but they are also precursors for a number of other important biomolecules (e.g. melanin, thyroxine, adrenaline). Excess amino acids in the diet (derived from protein digestion) are not stored but are broken down. The nitrogenous part is converted into urea for excretion while the carbon skeletons may be used in various ways. In principle, the carbon skeletons are ultimately oxidised to provide energy. However, it is more usual to speak of amino acids being ketogenic or glucogenic on the basis of whether their carbon skeletons are broken down via acetate (acetyl-CoA), and would therefore be capable of producing ketosis, or whether they might be converted into glucose via gluconeogenesis. Some amino acids are both ketogenic and glucogenic.

Humans can synthesize about half of the 20 amino acids needed for protein synthesis: the rest need to be supplied in the diet. Thus it is said that some amino acids are *essential* and some are *non-essential.* When proteins are being synthesized, an adequate supply of all 20 amino acids must be present. The system cannot 'wait' until an amino acid turns up and nor can it continue in the absence of one of the amino acids.

The body therefore needs to operate its amino acid economy very carefully. In the absence of a storage mechanism, it must bring together all 20 amino acids in the cell cytosol for protein biosynthesis, synthesizing those it can, if necessary, and obtaining the others in the diet. It must also have pathways for disposing of the amino acids. For an adult in nitrogen balance an amount of amino acids equal to the entire dietary intake is degraded each day. Thus, in the metabolic map there are a great number of pathways for dealing with amino acids (biosynthetic, degradative, transaminating, deaminating) and a correspondingly large number of enzymes. This gives a great potential for genetic diseases. A mutation affecting just one enzyme in just one amino acid metabolising pathway can have very severe or even fatal results.

There are indeed very many inherited diseases involving enzymes of amino acid metabolism (see Appendix 4). In the past they have been diagnosed by biochemical criteria, but in the future we may expect individuals at risk to be identified by prenatal diagnosis. Eventually we may see gene and enzyme-replacement therapy for treating diseased individuals in place of the dietary restriction treatments which are all that can be achieved at present.

The Problem

A male child, Simon V., had been born after a normal pregnancy but thereafter had failed to develop normally. When seen at age 10, he was unable to feed himself, could say only a few words and displayed involuntary limb movements. His IQ was 65. He had light blue eyes, light-coloured skin and his hair was white in places. His urine had a musty odour. Large amounts of an amino acid were present in the urine. In addition, a compound (compound I) was isolated from the urine in substantial amounts which, when incubated with NADH and lactate dehydrogenase (LDH), gave rise to compound II:

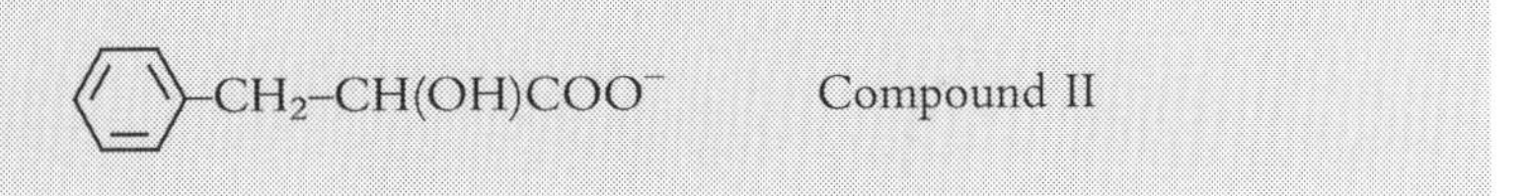

Questions

1. What techniques might be used to identify the unknown amino acid in the urine?
2. What is the structure of compound I and which class of reaction is involved in the formation of compound I from the amino acid that is present in the urine?
3. What is the metabolic block likely to be in the child? If it were necessary to confirm this experimentally, what procedure would you adopt?
4. What biochemical explanations can you provide for the light pigmentation and the abnormal development and mental retardation of the child?
5. If the condition had been diagnosed at birth, how could the child have been treated to ensure normal development?

Commentary

The unknown amino acid could be identified by paper electrophoresis, ion-exchange chromatography or high-performance liquid chromatography. A comparison with the 20 standard amino acids would be needed to identify the unknown. The pH at which the separation is performed is important and must be chosen with care.

Compound II is phenyllactate and is formed from compound I (phenylpyruvate) by reduction with NADH catalysed by LDH. The formation of phenylpyruvate from the unknown amino acid is an example of transamination, identifying the amino acid as phenylalanine. Transaminase reactions are of the following type:

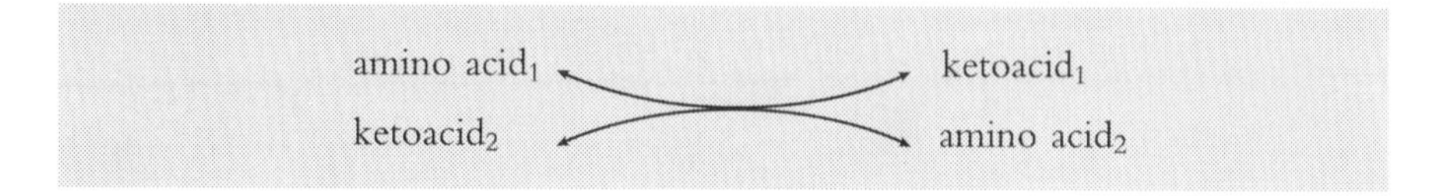

where an amino group ($-NH_2$) is transferred from an amino acid to a ketoacid using pyridoxal phosphate as cofactor. Urinary phenylpyruvate used to be detected by the ferric chloride test which produces a blue/green colour in the presence of the ketoacid; however, this test is not used these days for routine screening for this particular disorder because of its lack of sensitivity and specificity.

The accumulation of phenylalanine in blood (phenylalaninaemia) and its metabolites in the urine points to the genetic disorder known as phenylketonuria (PKU) in which the conversion of phenylalanine to tyrosine is the metabolic block. It is the most common inborn error of amino acid metabolism with an incidence among Caucasians of about 1 in 10 000 and even higher in those of Celtic origin (about 1:4 500). The reaction is catalysed by liver phenylalanine hydroxylase and is an example of a monooxygenase (or mixed-function oxygenase) reaction, in which one atom of oxygen appears in the product and the other in water. The cofactor tetrahydrobiopterin (BH_4) is required as a reductant in the reaction and is then reconverted to its active tetrahydro form through the action of a separate enzyme, dihydropteridine reductase (DHPR):

Phe hydroxylase:	$\text{Phe} + BH_4 + O_2 \rightarrow \text{Tyr} + H_2O + BH_2$
DHPR:	$BH_2 + \text{NADH} + H^+ \rightarrow BH_4 + \text{NAD}^+$

PKU is sufficiently common and potentially serious that all infants are tested for hyperphenylalaninaemia at about one week of age. Screening for PKU in the newborn child requires estimation of blood phenylalanine levels. Blood is taken by 'heel prick' on to filter paper for assay by the microbiological Guthrie test or high-performance liquid chromatography. However, if it were necessary to confirm the enzyme defect it would require direct determination of phenylalanine hydroxylase activity in the child. Phenylalanine hydroxylase is expressed only in liver and its assay *in vitro* requires a liver biopsy. However, liver biopsy is a stressful, invasive procedure and would not normally be performed in a case of PKU. Exposing the patient to a dose of phenylalanine and observing a rise in urinary metabolites is both unethical (phenylalanine causes brain damage in the developing child) and fails to identify the deficient enzyme directly.

Several factors probably contribute to the phenotypic characteristics of PKU, particularly the high blood levels of phenylalanine and some of its metabolites (phenylpyruvate and phenyllactate) and a deficiency of tyrosine. The light pigmentation is due to a lack of melanin which is synthesized from tyrosine via L-DOPA (dihydroxyphenylalanine) in a reaction catalysed by the enzyme tyrosinase. Tyrosine is also required for the synthesis in the brain of the catecholamine neurotransmitters, principally dopamine and noradrenaline:

$$\text{Tyr} \rightarrow \text{L-DOPA} \rightarrow \text{dopamine} \rightarrow \text{noradrenaline}$$

In this case, the conversion of tyrosine to L-DOPA is catalysed by tyrosine hydroxylase which is a monooxygenase like phenylalanine hydroxylase and also uses tetrahydrobiopterin as cofactor. Disturbances in neurotransmitter levels and homeostasis may also be involved in the abnormal brain development and mental retardation in the child. The exact mechanism involved is uncertain but there are several plausible explanations. Hyperphenylalaninaemia, rather than tyrosine deficiency, is probably the chief culprit since tyrosine supplementation alone is not an effective treatment for PKU. Early diagnosis and treatment is essential since there is estimated to be a loss of 5 IQ points for each ten-week delay in treatment.

Essentially the treatment is to prevent hyperphenylalaninaemia by giving a diet low in phenylalanine and adequate in other nutrients, including tyrosine. Phenylalanine cannot be eliminated completely from the diet since it is an essential amino acid. Since phenylalanine occurs in nearly all proteins, artificial dietary formulations are required. Recently, inclusion in the diet of a mixture of neutral amino acids (valine, isoleucine, leucine) has been suggested as a beneficial supplement in the treatment of PKU. The neutral amino acids are transported into the brain by the same carrier protein that transports phenylalanine. Thus they can competitively inhibit the entry of phenylalanine into brain.

It has been common practice to cease dietary restriction when brain growth and development are thought to be complete but more recent evidence suggests it may be advisable to continue treatment throughout childhood and adolescence, and perhaps for life. It is essential for pregnant women with PKU to control their blood phenylalanine levels to prevent damage to the foetus, so-called 'maternal PKU'.

Hyperphenylalaninaemia need not solely be due to a defect in phenylalanine hydroxylase. Reduced levels of the BH_4 cofactor can give rise to an atypical form of PKU, often referred to as 'malignant PKU' because the neurological deficits in such patients are not prevented by dietary restriction alone. This is because BH_4 is also required as a cofactor for tyrosine hydroxylase (see above) and for tryptophan hydroxylase, which is involved in the biosynthesis of serotonin. BH_4 is also a cofactor for nitric oxide synthase which catalyses the formation of the signal molecule nitric oxide from the amino acid arginine in brain and other tissues. A deficiency of BH_4 can arise from a defect in any of the enzymes involved in its synthesis from GTP, or in DHPR. The cofactor can be measured in blood by a bioassay involving growth of the protozoan, *Crithidia fasciculata*. PKU patients with biopterin deficits need therapy which combines restriction of phenylalanine intake with maintenance of neurotransmitter homeostasis by supplementation of the diet with BH_4, L-DOPA and 5-hydroxytryptophan.

The human phenylalanine hydroxylase gene has now been cloned allowing prenatal diagnosis in PKU families. Appropriate oligonucleotide probes have been developed which can be used to distinguish between normal and mutant alleles in the genome. However, a number of different mutations are associated with PKU, so screening a population would require the use of several different probes. Within an individual family a single probe may be diagnostic.

Further Questions

1. How does the Guthrie test work?
2. Why should there be concern about phenylketonurics, and even heterozygous PKU carriers, using the artificial sweetener, aspartame (*N*-aspartylphenylalanine methylester)?
3. You are the Chief Medical Officer of Health in a country where it is proposed to screen all newborn infants for a potentially serious genetic condition (incidence 1:15 000 live births) by using a blood test (cost £5) which, in trials, identified 90% of affected children and scored 1% of normal children as 'false positives'. Discuss the advantages and disadvantages of the proposal with your Minister of Health. What other strategies could be adopted?
4. Two competing pathways for phenylalanine metabolism are hydroxylation and transamination. The data in Figure 19.1 show the relative rates of hydroxylation and transamination as a function of phenylalanine concentration. Use the data in this figure to explain why PKU is characterised by excessive urinary excretion of secondary metabolites of phenylalanine such as phenylpyruvate arising from the transamination pathway.

Figure 19.1. Relative rates of phenylalanine hydroxylation and transamination.

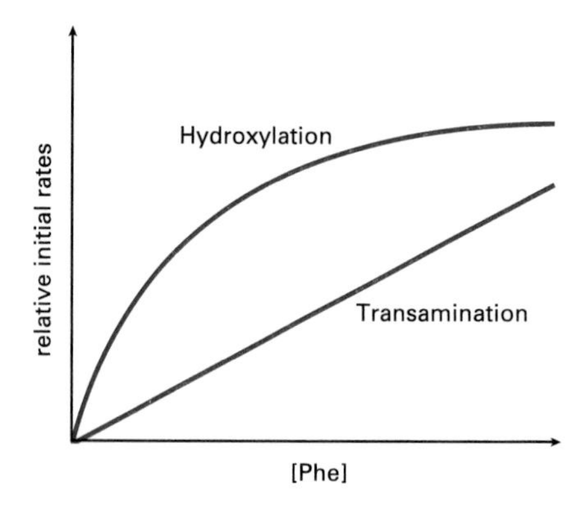

Figure 19.2. Pedigree analysis of Simon V.'s family.

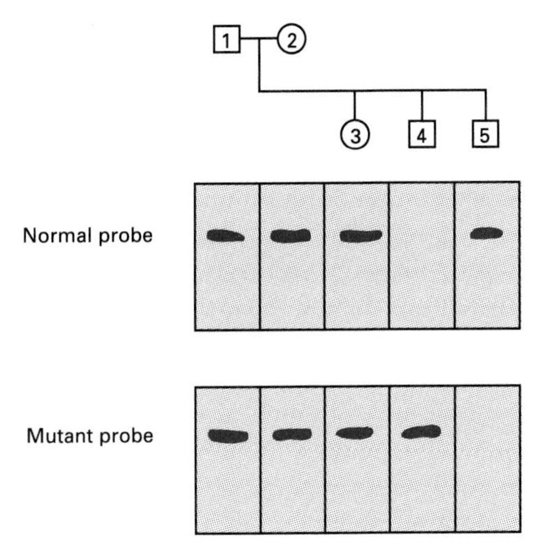

Figure 19.3. Analysis of phenylalanine hydroxylase and its mRNA expressed in transfected cells in culture.

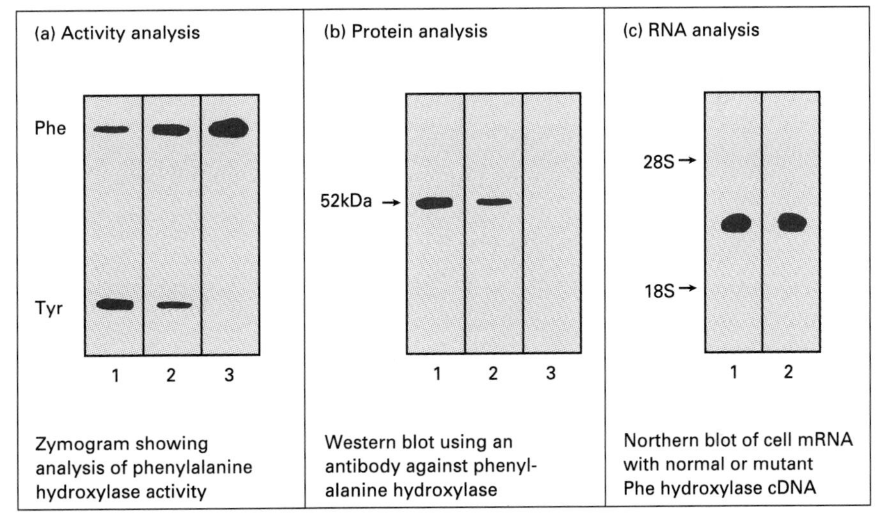

5. Figure 19.2 shows the pedigree analysis of the patient's family. Simon (4) has an older unaffected sister (3) and a newborn brother (5). Both his father (1) and his mother (2) are unaffected. Southern blotting of leukocyte DNA from members of the family (lanes correspond to the numbers used in the pedigree) is shown with normal probe and mutant probe. Deduce the genotype of each member of the family. What type of inheritance is indicated? Should Simon's brother be treated for PKU? What should a genetic counsellor tell Simon's sister when she marries?

6. In a further study (Figure 19.3), cells in culture transfected with the normal or the mutant phenylalanine hydroxylase cDNAs were subjected to biochemical analysis. Panel (a) shows the assay of phenylalanine hydroxylase activity purified from normal liver (track 1), extracts from cells transfected with normal phenylalanine hydroxylase cDNA (track 2), and from cells transfected with mutant cDNA (track 3). Panel (b)

shows a Western blot analysis of the hydroxylase in which the protein (M_r 52 kDa) has been detected by an antibody to the enzyme after SDS–PAGE and transfer to a nitrocellulose sheet. Track 1 reveals the hydroxylase from a normal human liver extract; track 2 is an extract from cells transfected with normal hydroxylase cDNA; track 3 is an extract from cells transfected with mutant hydroxylase cDNA. Equal amounts of cell protein were analysed in each lane. Panel (c) shows a Northern blot analysis of the hydroxylase mRNA content of cells transfected with of normal (track 1) or mutant (track 2) hydroxylase cDNA (equal amounts of RNA were analysed). What can you deduce about the expression of the mutant phenylalanine hydroxylase gene?

7. Calculate the proportion of PKU heterozygotes in the population (PKU incidence is 1:10 000 live births).

Connections

- This question is about one of the best characterised metabolic diseases. Use it to check that you understand the outline of amino acid metabolism, that is, why the body carries out all the various interconversions. Phenylalanine is an essential amino acid, but what about tyrosine? And what implication has this for a diet designed for PKU patients?
- The reactions by which phenylalanine is converted to tyrosine are of the oxygenase type. What other reactions (monooxygenase, dioxygenase) of this type have you encountered?
- Can you describe the rest of the pathway of phenylalanine degradation and do you understand how DOPA, adrenaline, noradrenaline, melanin and thyroxine are formed?
- Use this question to revise how the prenatal diagnosis of genetic disease is carried out. What are the relative advantages of amniocentesis and chorionic villus sampling as ways of obtaining samples for testing? What other genetic diseases are you likely to encounter regularly during your medical practice (see Appendix 4)? (See also Problems 1, 5, 7, 8, 10–12, 15, 18, 20 and 22.)
- Can you remember how the amino groups of amino acids are dealt with when the body needs to get rid of them? Which amino acid directly generates urea? (See also Problem 5.)
- How much protein does a normal individual really need per day in his/her diet? It is salutary to compare the diet of people in the Northern hemisphere with that of those in the Southern hemisphere. To what dangers are vegans exposing themselves? (See also Problems 2 and 6.)
- Untreated diabetics are thin because they are breaking down their muscle mass. Can you explain this and its relevance to amino acid metabolism? Does it also account for diabetic ketosis? (See also Problems 2 and 18.)
- There are links in this question with parkinsonism. Do you know what Parkinson's disease is and how it may be treated? (See also Problem 21.)

References

DiLella A G, Huang W-M and Woo S L C (1988) Screening for phenylketonuria mutations by DNA amplification with the polymerase chain reaction. *Lancet* 1 (March 5), 497–9.

Levy H L (1989) Molecular genetics of PKU and its implications. *American Journal of Human Genetics* **45**, 667–70.

Matalon R and Michals K (1991) Phenylketonuria: screening, treatment and maternal PKU. *Clinical Biochemistry* **24**, 337–42.

Scriver C R and Clow C L (1988) Avoiding phenylketonuria: why parents seek prenatal diagnosis. *Journal of Pediatrics* **113**, 495–7.

de Swiet M (1992) Medical disorders in pregnancy. *Current Opinion in Obstetrics & Gynecology* **4**, 28–36.

Verma I M (1990) Gene therapy. *Scientific American* **263**(5), 34–41.

Woo S L (1989) Molecular basis and population genetics of phenylketonuria. *Biochemistry* **28**, 1–7. Short review article.

20 WORRIED ABOUT HEART ATTACK

Introduction

In Western societies cardiovascular disease, including ischaemic heart disease, is the single most important cause of death (Figure 20.1). In fact cardiovascular disease claims almost as many victims as all the other causes of death put together. Ischaemic heart disease shows a wide geographical variation, with Scotland and Finland at the head of the world league table in the 1980s (Table 20.1). The aetiology of ischaemic heart disease is complex and involves both genetic predisposition and environmental risk factors. The latter include dietary factors, smoking, lack of exercise and stress.

One of the indicators of susceptibility to cardiovascular disease is hypercholesterolaemia. In view of this, a considerable amount of research over many years has gone into establishing the nature of the link between plasma cholesterol and arterial disease.

Cholesterol was first isolated from gallstones ('chole' is Greek for bile) but is now recognised as an essential component of all cell membranes. Cells can synthesize cholesterol *de novo* from acetyl-CoA or they can import it from the blood. Because of its hydrophobic nature, cholesterol, like other lipids, has to be transported in the form of lipoprotein complexes. Low-density lipoprotein (LDL) is particularly rich in cholesterol, and especially cholesteryl esters. Many cells are able to take up LDL particles and use them as an exogenous source of cholesterol, thus decreasing their need for endogenously-synthesized cholesterol.

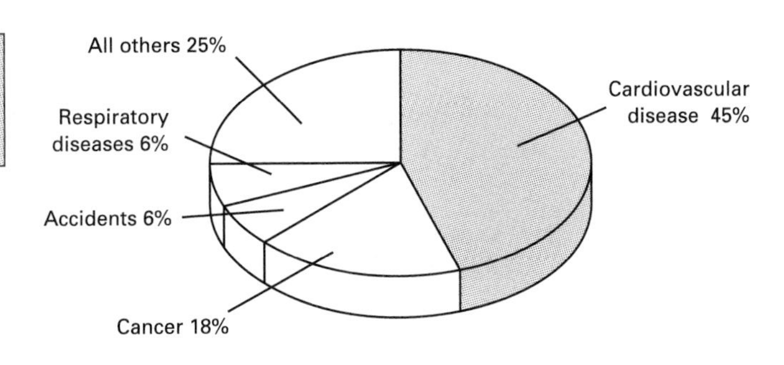

Figure 20.1. Major causes of death in Western society.

Table 20.1 Death rates from ischaemic heart disease in males (age-standardised per 100 000 population). Taken from WHO Health Statistics Annual, 1984.

Country	Rate	Country	Rate
Scotland, Finland	435	Norway	300
Ireland	385	Germany, Netherlands	255
New Zealand, Sweden	380	Austria, Bulgaria	245
Czechoslovakia	370	Belgium	200
England & Wales	360	Romania	190
Australia, Denmark	355	Italy	180
USA	350	Greece	120
Hungary	340	France	110
Canada	325	Japan	60

The Problem

A 38-year-old self-employed business man, Brian C., went to his physician concerned about his health. His business was doing badly and he was working long hours. He ate irregularly, smoked heavily and drank substantial quantities of alcohol. He had noticed a number of raised areas (xanthomata) on his hands and legs (associated with the extensor and Achilles tendons, respectively). His doctor found by questioning that some members of his family had died relatively young from heart attacks.

A fasting blood sample was assayed for plasma lipids (Table 20.2). The plasma proteins were separated by agarose gel electrophoresis and were then visualised by staining with a lipophilic dye (Figure 20.2). The cholesterol content of each of the main plasma components identified by electrophoresis was calculated (Table 20.3).

Fibroblasts were cultured from Brian's skin. Their ability to bind radiolabelled LDL was compared with that of control fibroblasts (Figure 20.3a). After incubation with LDL their 3-hydroxy-3-methylglutaryl-CoA (HMG-CoA) reductase activity was also measured (Figure 20.3b).

In view of these results and the case presentation, the patient's LDL receptor genes were examined by Southern blotting. A Southern blot (named after Ed Southern, its inventor) is a common way of exploring gene structure. Restriction endonucleases, which recognise specific nucleotide sequences in DNA, are used to cut the gene

Table 20.2. Plasma lipid analysis.

Lipid	Patient	Normal
Triglyceride (mM)	1.7	0.6–3.2
Cholesterol (mM)	9.5	3.7–6.8

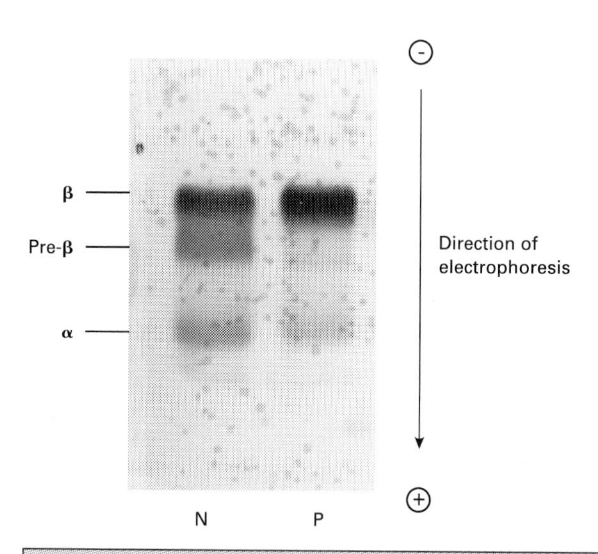

Figure 20.2. Agarose gel electrophoresis of plasma proteins (N, normal; P, patient). The gel was stained with a lipophilic dye and the protein bands are designated by their electrophoretic mobilities relative to the plasma α and β globulins. Photograph kindly supplied by Dr Anne Soutar, Medical Research Council Lipoprotein Team, Hammersmith Hospital, London.

Table 20.3. Cholesterol analysis of plasma electrophoretic components (values mM with respect to plasma).

Subject	β	Pre-β	α
Normal	3.18	0.41	1.37
Patient	7.70	0.70	1.14

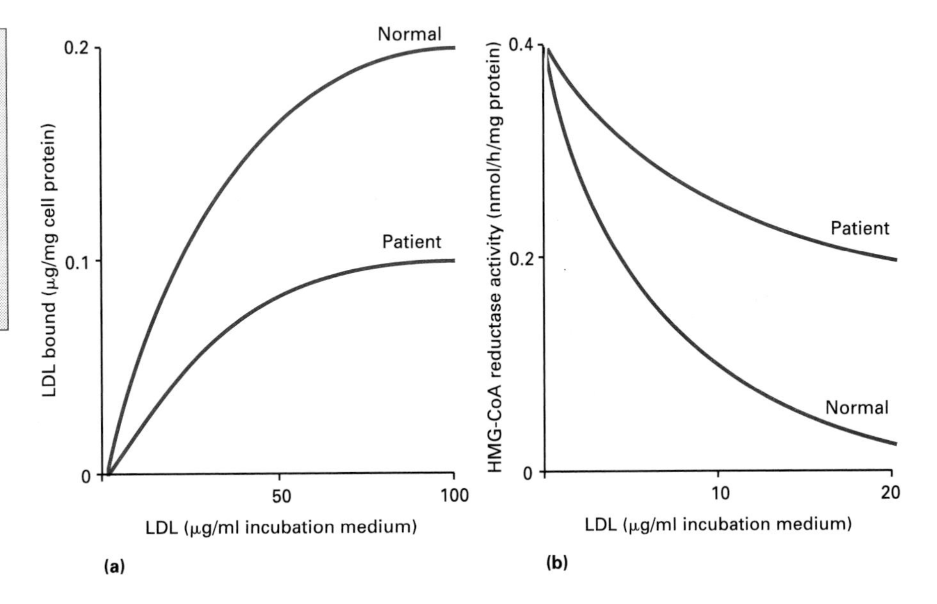

Figure 20.3. Effects of LDL concentration on binding of radiolabelled LDL to cultured fibroblasts (a) and on the activity of HMG-Co A reductase (b). Adapted from Brown and Goldstein (1974 and 1979).

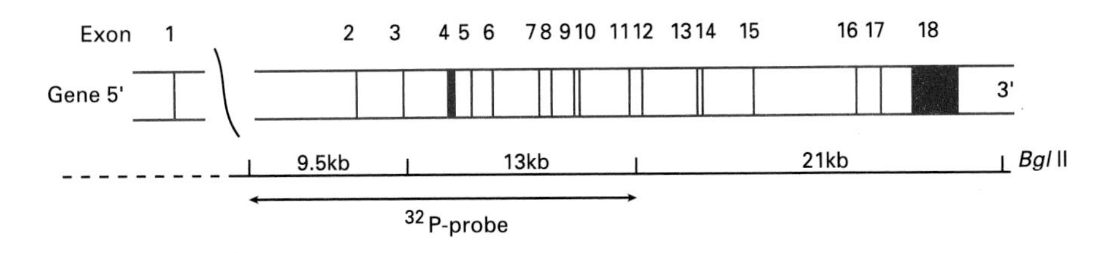

Figure 20.4. Diagram of the human LDL receptor gene and its *Bgl* II restriction map. The human LDL receptor gene is very large and has 18 exons. The cleavage sites for the restriction enzyme *Bgl* II are shown together with the gene region which is complementary to the probe used in the Southern blot. Sizes of restriction fragments are given in kilobase-pairs (kb). Based on a diagram from Dr Anne Soutar, Medical Research Council Lipoprotein Team, Hammersmith Hospital, London.

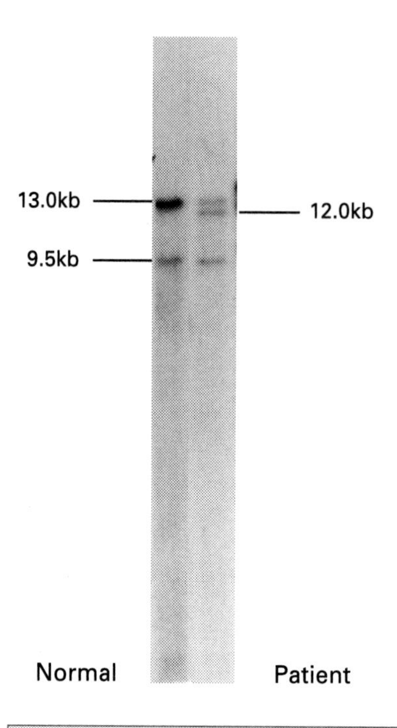

Figure 20.5. Southern blotting of the LDL receptor gene. DNA samples from the patient and a normal subject were restriction digested with *Bgl* II. The restriction fragments were separated by electrophoresis and blotted onto nitrocellulose. The nitrocellulose blot was then hybridised with the radiolabelled probe shown in Figure 20.4 and autoradiographed. Photograph kindly supplied by Dr Anne Soutar, Medical Research Council Lipoprotein Team, Hammersmith Hospital, London.

into defined fragments. These restriction fragments are separated according to their size by gel electrophoresis and then transferred from the gel onto a sheet of nitrocellulose or nylon by a blotting procedure in such a way that the blot represents a replica of the electrophoretic pattern of separated fragments in the gel. The presence of a particular DNA sequence in any of the restriction fragments is then established by screening the blot with a radiolabelled nucleic acid probe specific for the gene in question. Fragments which contain sequences complementary to the probe will base-pair (hybridise) to the probe. Hybridised probe is detected by autoradiography of the blot. Figure 20.4 shows a diagram of the LDL receptor gene and its restriction map for the enzyme *Bgl* II. Figure 20.5 is the autoradiograph of the Southern blot of Brian's DNA and the DNA of a normal subject restricted with *Bgl* II and probed with the [^{32}P]probe shown in Figure 20.4.

Brian was given general advice about his life-style and put on a course of oral cholestyramine (a synthetic anion-exchange compound which is not absorbed from the gut) and lovastatin (a competitive inhibitor of HMG-CoA reductase). Four months later, his fasting plasma cholesterol had fallen by 30%.

Questions

1. In what way is the patient's plasma lipid analysis abnormal?
2. What are the identities of the plasma components revealed by electrophoresis? What is the abnormality in the patient and can this explain the lipid analyses?
3. What type of mechanism is likely to be involved in the uptake of LDL by cells (Figure 20.3a)? What do you suppose is the fate of the LDL inside the cell?
4. What is the significance of HMG-CoA reductase in cholesterol biosynthesis? In the experiment shown in Figure 20.3b, can you explain why LDL lowers the HMG-CoA reductase activity in normal cells?
5. What metabolic defect in the patient is suggested by the data in Figure 20.3?
6. What information is there in this case that prompted the suspicion of a genetic defect? What type of LDL gene mutation is revealed by the Southern blot (Figure 20.5)? How can this be the basis of the metabolic defect you have identified?
7. The liver uses cholesterol to synthesise bile salts (glycocholate and taurocholate) which aid the digestion of fats and their absorption from the gut. In the process these bile salts are re-absorbed. What is the rationale behind the use of cholestyramine in Brian C.'s treatment?
8. Why was lovastatin also used?
9. If you had been Brian C.'s doctor what recommendations would you have made concerning his life-style?
10. Brian C. has several siblings and a teenage child. Is there any information in this problem which suggests that you should extend your investigation to these family members? What tests would you order?

Commentary

Brian C.'s plasma lipid analysis revealed hypercholesterolaemia, suggesting that he is at risk of cardiovascular disease. Brian suffers from familial hypercholesterolaemia (FH), also called hyperlipoproteinaemia type IIa.

The cholesterol in Brian's blood has to be carried in the form of lipoprotein complexes of which there are several different types in plasma. Each consists of a non-polar lipid core surrounded by a hydrophilic shell of apoproteins and polar lipids. They differ in the types and proportions of these components and this determines their relative densities. Experimentally, these differences can be exploited during ultracentrifugation to separate the lipoproteins into zones according to their buoyant density. Very low-density lipoprotein (VLDL), as its name implies, has a relatively low-density because its lipid:protein ratio is high. High-density lipoprotein (HDL), on the other hand, is characterised by a lower lipid:protein ratio and is correspondingly more dense. The composition and density of low-density lipoprotein (LDL) are intermediate between those of VLDL and HDL. Lipoproteins may also be separated by electrophoresis as shown in Figure 20.2. VLDL, LDL and HDL are the principal lipoprotein constituents of the pre-β-, β- and α-electrophoretic components, respectively. LDL is the principal carrier of plasma cholesterol (mainly in the form of cholesteryl ester) (see Table 20.3). Examination of the data in Figure 20.2 and Table 20.3 shows that Brian has elevated plasma LDL and this provides an explanation for his hypercholesterolaemia.

Many cells can express surface receptors for LDL. These receptors recognise the apolipoprotein in LDL, bind the LDL and then cluster together in clathrin-coated pits in the plasma membrane. The membrane in the vicinity of the pit invaginates to internalise the LDL receptors along with their bound LDL in a process known as endocytosis. The endocytic vesicles eventually fuse with lysosomes and the LDL is degraded to release its cholesterol cargo. This endogenously-acquired cholesterol then suppresses *de novo* cholesterol biosynthesis from acetyl CoA by a negative-feedback mechanism in which the activity of the enzyme catalysing the committed step in cholesterol biosynthesis, HMG-CoA reductase, is decreased.

The LDL binding curve for Brian's fibroblasts reaches a plateau at about 50% of the normal value, but the LDL concentration for half-maximal binding is unaltered (Figure 20.3a). This implies that Brian's cells possess functionally normal LDL receptors, but only half the complement of normal cells. As a consequence, they internalise less cholesterol (as LDL) and there is less suppression of their HMG-CoA reductase (Figure 20.3b). This explains why Brian's plasma LDL is higher than normal; less LDL is being removed from the circulation.

A familial disorder can be suspected since some of his family had died from heart disease at a relatively young age. The apparent decrease in his LDL receptors suggested that his LDL receptor genes might be defective. The restriction map (Figure 20.4) shows that the probe used in the Southern blot should hybridise to two *Bgl* II fragments (13.0 and 9.5 kb) from the LDL receptor gene. The autoradiograph (Figure 20.5) shows bands corresponding to these two fragments from the normal subject's DNA. These are also present in Brian's sample but there is an additional unexpected band corresponding to a restriction fragment of 12.0 kb. Furthermore, the intensity of the autoradiographic signal from the 13.0 kb fragment is weaker than normal. These results suggest that Brian is heterozygous for a mutation in one of the diploid copies of his LDL receptor gene which has deleted 1.0 kb from the 13.0 kb fragment, replacing it with the shorter 12.0 kb fragment. We can discount an alternative explanation – that the mutated gene has gained a new *Bgl* II cleavage site, now allowing the 13.0 kb fragment to be cut into two fragments (12.0 kb and 1.0 kb) – by the absence of a 1.0 kb band on the Southern blot. Although the Southern blot does not indicate the precise position of the deletion in the gene, the 13.0 kb region encompasses several of the gene's exons so it is likely that the mutated allele would produce a functionally abnormal or unstable LDL receptor. This would explain why Brian's cells appear to have only half of the normal complement of functional LDL receptors (Figure 20.3a).

Single-gene disorders are often recessive, i.e. heterozygous individuals are unaffected. However, Brian is clearly heterozygous for the LDL receptor gene mutation and yet he has hypercholesterolaemia, so we may surmise that FH is an autosomal dominant condition. With the evidence before us, it would be sensible to screen the immediate members of his family for hypercholesterolaemia and the mutant allele.

In the long term, elevated plasma LDL is thought to result in deposition of cholesterol at certain sites. These include the xanthomata characteristic of FH. Much more serious, however, is the development of atheromatous plaques in the walls of blood vessels. These may become large enough to impede circulation (Figure 20.6), especially to the heart.

Although FH has quite a high incidence (about 1 in 500 of the population), most hypercholesterolaemic individuals do not suffer from FH. Rather they have a complex genetic background which disposes them to ischaemic heart disease in the face of various detrimental environmental factors, such as smoking, alcohol, stress, lack of exercise and diets low in fibre but rich in animal fats. Nevertheless, a study of FH has provided some insights into the causation of cardiovascular disease in the general population. Bearing in mind the predisposing risk factors, Brian and other affected family members should be counselled about their life-styles.

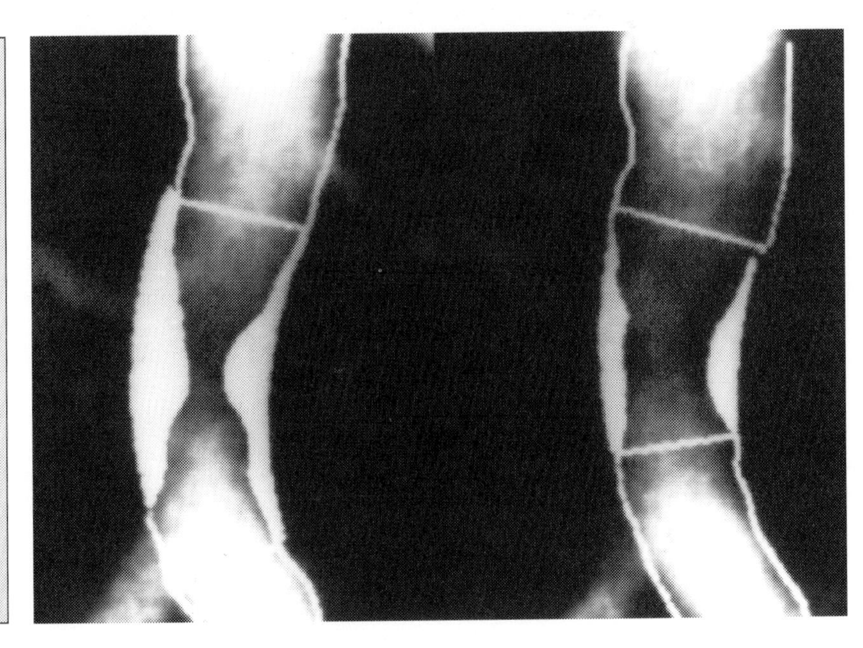

Figure 20.6. Computer-assisted reconstructions of angiograms showing a stenotic lesion in proximal left circumflex artery before (left) and after (right) treatment to reduce plasma LDL levels. Photograph kindly supplied by Dr Gilbert Thompson, Medical Research Council Lipoprotein Team, Hammersmith Hospital, London, and reproduced from Barbir *et al.* (1989) *British Medical Journal* **298**, 132.

The rationale of Brian's drug treatment is to increase the cellular demand for cholesterol in the form of LDL. If more LDL can be taken up, there will then be less in circulation to form atheromata. This is achieved by a two-fold approach:

(i) Lovastatin suppresses *de novo* cholesterol synthesis by inhibiting HMG-CoA reductase and forces cells to use exogenous cholesterol.
(ii) Cholestyramine prevents re-absorption of bile salts from the gut for use as a source of exogenous cholesterol.

This forces cells to obtain their cholesterol by endocytosis of LDL with a consequent fall in the plasma LDL level. In the long term this treatment should both diminish the tendency for atherosclerotic plaques to build up in the blood vessels and allow the regression of existing lesions as shown in Figure 20.6.

Further Questions

1. What lipoproteins would be revealed on electrophoresis of plasma taken from a normal person soon after a fatty meal?
2. A different type of hyperlipoproteinaemia, type I, can be due to a deficiency in lipoprotein lipase. What symptoms would you predict and what tests could be done to distinguish it from FH?
3. Liver damage often results in abnormalities in lipoprotein metabolism. What changes would you expect, and why? (See also Problem 17.)
4. In the case described here, the inherited hypercholesterolaemia was due to a faulty LDL receptor gene. From the information about LDL metabolism in this problem, can you suggest other ways in which inherited hypercholesterolaemia could arise?
5. Would the treatment given to Brian C. be effective for a homozygous patient?
6. How is the atheromatous plaque thought to arise?
7. Do the data in this case give any clues about how a diet high in fibre might protect against elevated plasma cholesterol?
8. How would you show that FH is an autosomal dominant condition?

Connections

- Use this problem to revise your knowledge of fat digestion and absorption. Check that you know how lipids are transported, how fat is deposited in adipose tissue and subsequently mobilised.
- This problem involved genetic screening, the use of restriction enzymes and hybridisation probes. For screening it is not necessary to obtain DNA directly from

the affected organ. Check that you understand why; suggest suitable alternative sources. Are you clear about how a hybridisation probe works and how you would design a probe and the hybridisation protocol? (See also Problem 8.)

- The diagram in Figure 20.4 shows that the LDL receptor gene is composed of exons and introns. Do you understand these terms?
- Brian C.'s treatment involved using a drug as an inhibitor of a key regulatory enzyme. Use this to review the different types of enzyme inhibition and ensure that you can distinguish between reversible and irreversible inhibition, competitive and non-competitive inhibition. List other examples of drugs which act as enzyme inhibitors. (See also Problems 16, 17 and 21.)
- Review the structure and function of cell membranes. Think of the various processes whereby material may enter cells. Distinguish between active and passive transport. (See also Problem 13.)
- The endocytosis of LDL involved recognition by a specific cell surface receptor. Think of the varied roles they can perform. Summarize the roles of receptors which are located inside the cell. Make a list of the biochemical properties which you would expect of receptors. (See Problems 9, 15, 18 and 22.)

References

Breslow J L (1989) Genetic basis of lipoprotein disorders. *Journal of Clinical Investigation* **84**, 373–80.

Brown M S and Goldstein J L (1974) Expression of the familial hypercholesterolemia gene in heterozygotes: mechanism for a dominant disorder in man. *Science* **185**, 61–3. Original report.

Brown M S and Goldstein J L (1979) Receptor-mediated endocytosis: insights from the lipoprotein receptor system. *Proceedings of the National Academy of Sciences U.S.A.* **76**, 3330–7. Original report.

Brown M S and Goldstein J L (1990) Scavenging for receptors. *Nature* **343**, 508–9. *News and views* article.

Humphries S E (1992) The application of molecular biology techniques to the diagnosis of hyperlipidaemia and other risk factors for cardiovascular disease. *Journal of the International Federation of Clinical Chemistry* **4**, 208-18.

Krieger M (1992) Molecular flypaper and atherosclerosis: structure and function of the macrophage scavenger receptor. *Trends in Biochemical Sciences* **17**, 141–6.

Ross R (1993) The pathogenesis of atherosclerosis: a perspective for the 1990s. *Nature* **362**, 801–9.

Sacks F M and Willett W W (1991) More on chewing the fat: The good fat and the good cholesterol. *New England Journal of Medicine* **325**, 1740–1.

Soutar A K (1993) Investigation and diagnosis of disorders of lipoprotein metabolism. In *Molecular Biology in Clinical Research and Diagnosis* (Walker M and Rapley R, eds). Blackwell Scientific Publications, Oxford.

Thompson G R (1989) *A Handbook of Lipoproteinaemia.* Current Science Ltd, London.

21 DESIGNER DRUG

Introduction

Signalling between two nerve cells is a chemical event mediated by one or more of a wide variety of neurotransmitters. The monoamine transmitters, which are important in the control of mood, movement and the integration of thought processes, include the catecholamine family (principally dopamine and noradrenaline) and the indoleamine, serotonin (5-hydroxytryptamine, 5-HT). These transmitters are synthesized in a series of steps from their precursor amino acids tyrosine and tryptophan, respectively. The amino acids, but not the amine products, are able to cross the blood–brain barrier readily and enter neurones where they are converted to their end products in a series of enzymic steps and then packaged into storage granules (synaptic vesicles) in the nerve ending. On stimulation of the nerve cell, the packages of neurotransmitter are released into the synaptic cleft and interact with specific receptors on the post-synaptic cell membrane. Synaptic transmission is rapid and brief, the transmitter being quickly removed to terminate the process. Several inactivation mechanisms for neurotransmitters have been identified. Some (e.g. acetylcholine, neuropeptides) are destroyed in the synaptic cleft by enzymes located in the pre- or post-synaptic membranes. Alternatively, as for the monoamines and γ-aminobutyrate (GABA), the neurotransmitters are removed by being pumped back into neurones or their surrounding glial cells. This uptake mechanism is usually Na^+-dependent and involves a family of membrane transport proteins which have similar topology to the glucose transporter (Figure 21.1).

In the treatment of conditions such as depression, anxiety and schizophrenia, psychiatric medicine makes use of numerous drugs that interfere with one or more steps in the neurotransmitter cycle (Figure 21.2). Drugs of abuse such as heroin, cocaine, 'ecstasy' and LSD also interfere dramatically with the process of neurotransmission. In the problem described here, a synthetic heroin product injected by a drug addict produced a severe and irreversible neurological condition.

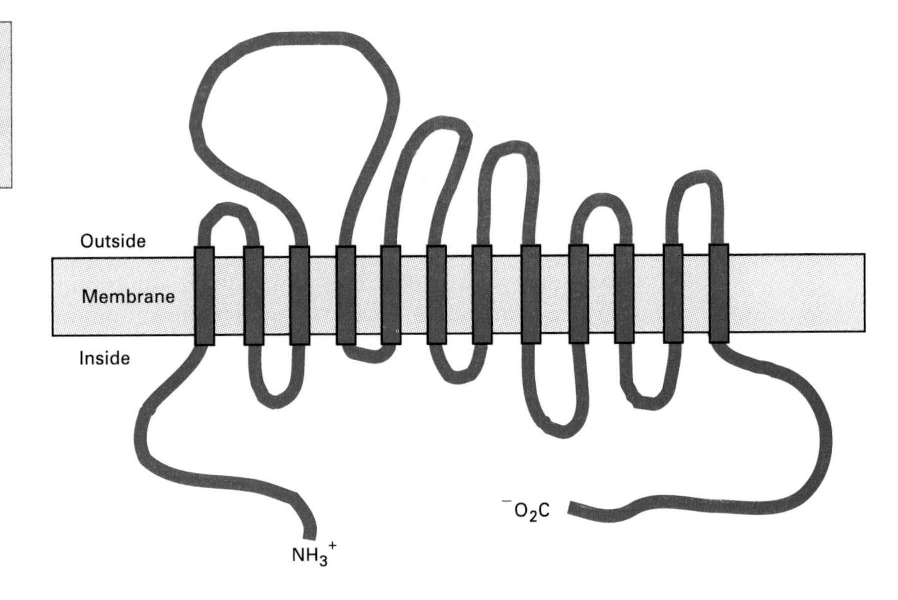

Figure 21.1. Predicted topology of a plasma membrane neurotransmitter transport protein.

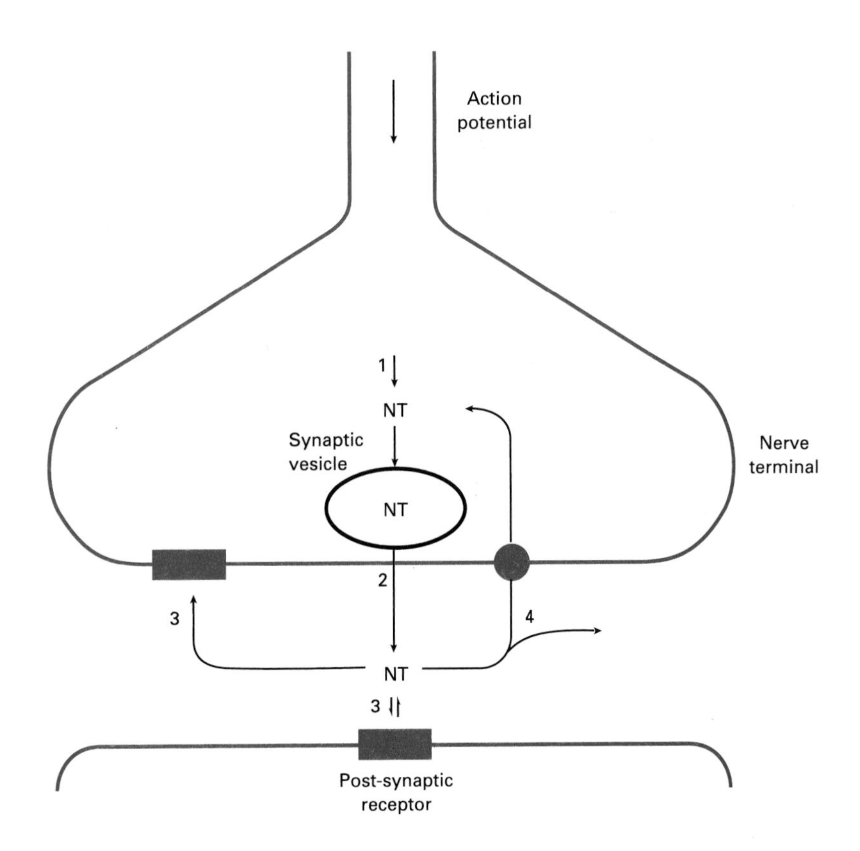

Figure 21.2. The neurotransmitter cycle and possible sites of drug action. 1, biosynthesis and storage in synaptic vesicles; 2, release of neurotransmitter (NT); 3, interaction of neurotransmitter with pre-synaptic or post-synaptic receptor; 4, termination of neurotransmitter action by uptake and/or metabolism.

The Problem

A young Californian student, Randy B., aged 26, was referred to a neurological department after reporting severe reactions to a synthetic heroin derivative (a 'designer drug') purchased illegally. The patient had a previous history of drug abuse. He initially suffered visual hallucinations, jerking of limbs and some stiffness. Within a week of starting the new drug he experienced a generalised slowing and difficulty in movement. The symptoms continued to worsen after discontinuing the drug, producing a near total immobility, a flexed posture, occasional tremor, difficulty in speaking and a marked reduction in rate of blinking. Randy responded to therapy with a combination of L-dihydroxyphenylalanine (L-DOPA) and carbidopa (an aromatic amino acid decarboxylase inhibitor). The inclusion of bromocriptine (a dopamine agonist) in the treatment further improved the response. When medication was subsequently stopped for a short period, the patient returned to his original state of rigidity and immobility. A positron emission tomography (PET) scan of Randy's brain was carried out after injecting him with 6-fluorodopa containing the positron-emitting isotope ^{18}F. Twelve sequential scans were performed and two adjacent axial scans through his striatum are shown in Figure 21.3 and compared with scans of a normal subject.

Chemical analysis of the designer drug that Randy had taken revealed that it was significantly contaminated with the compound 1-methyl-4-phenyl-1,2,5,6-tetrahydropyridine (MPTP). He subsequently died of a heroin overdose some two years after the initial onset of the neurological symptoms.

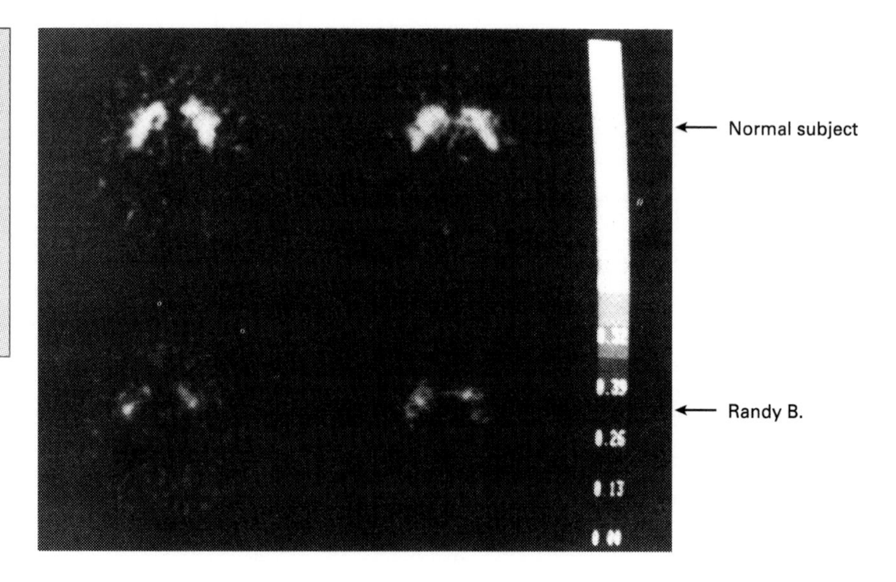

Figure 21.3. Fluorodopa PET brain scans of normal subject and Randy B. Scale on right represents radioactivity in arbitrary units, increasing from black to white. Reproduced from Calne *et al.* (1985) *Nature* 317, 246–8.

Questions

1. How and where are catecholamines synthesized?
2. What relatively common neurological condition, usually seen in older people, does the drug-induced disability closely resemble?
3. What information can be provided by a PET scan? What specific deficits might you expect to see *post mortem* in the brain of the patient?
4. Why should the administration of L-DOPA be more effective than its precursor, or its product? Why should the efficacy be improved by the inclusion of carbidopa and bromocriptine?
5. In experimental animals the onset of MPTP neurotoxicity can be prevented by pretreatment with deprenyl, an inhibitor of one form of monoamine oxidase. The effects of the oxidation product MPP^+ were therefore compared on the oxidation of glutamate/malate or succinate by rat brain mitochondria (Figure 21.4). MPTP itself had no effect on the oxidation of either substrate. What do these data suggest about possible steps in the toxicity of MPTP and its site of action?
6. For what clinical purpose are monoamine oxidase inhibitors used?

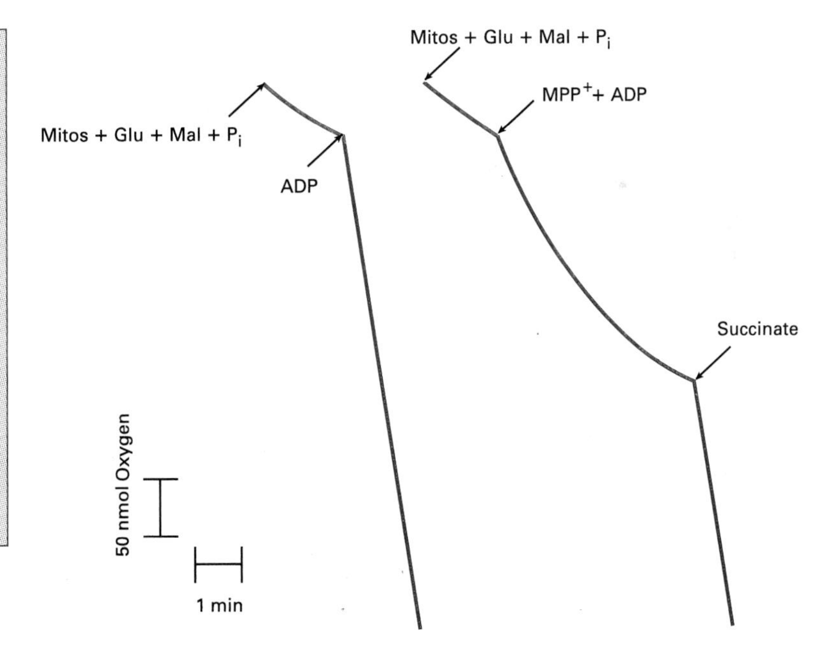

Figure 21.4. Oxygen electrode trace of the oxidation of glutamate/malate or succinate by isolated mitochondria in the presence or absence of MPP^+. Glu = glutamate; Mal = malate; P_i = inorganic phosphate; Mitos = mitochondria. Unpublished data generously made available by K F Tipton and G Davey (Trinity College, Dublin, Ireland).

Commentary

The pathway of catecholamine biosynthesis from tyrosine is shown in Figure 21.5. The principal sites of synthesis are the adrenal medulla and the brain.

The contaminant in the illicit opiate drug induced a syndrome closely resembling Parkinson's disease (PD) in the patient. This condition was first described by James Parkinson in 1817 and was referred to as the 'shaking palsy'. It is one of the commonest neurodegenerative diseases and is characterised by decreased movement (akinesia), resting tremor and rigidity. The disease normally begins after the age of 40 and usually progresses slowly with increasing disability. In the UK alone some 120 000 people suffer from Parkinson's disease.

PET allows the possibility of high-resolution anatomical localisation, within living tissues in conscious patients, of the distribution, metabolism or function of molecules present in minute amounts. A positron is a high-energy positive electron emitted from certain radioisotopes which are normally produced in a cyclotron. Positron-emitting isotopes have short half-lives and can be produced with high specific activities so they can be used in tracer amounts in patients, providing very high sensitivity. A variety of such positron emitters is used in medical research, depending on the particular application, e.g. ^{11}C, ^{13}N, ^{15}O, or ^{18}F as in this particular problem. The positron, on delivery to the target tissue, moves a short distance before it collides with an electron and is annihilated, resulting in the production of two photons of identical energy emitted virtually at 180° to each other. This allows for coincidence counting by cameras sited on opposite sides of the tissue and the emission events can be quantified precisely.

DOPA can be tagged directly with $^{18}F_2$ gas and is converted *in situ* in the brain to fluorodopamine which can then be visualised. Pathologically, PD is characterised by degeneration of dopamine neurones which have their cell bodies in a small nucleus in the brainstem known as the substantia nigra. These dopamine-containing nerve cells have long axons that terminate in the corpus striatum, an area of the brain concerned with the smooth co-ordination of limb movements. At least 70% of dopamine neurones in this nigrostriatal pathway must be lost before any clinical symptoms of PD appear. The PET scan allows the direct visualisation of dopamine and its metabolites in nerve endings in the living brain. The normal subject shows an accumulation of the

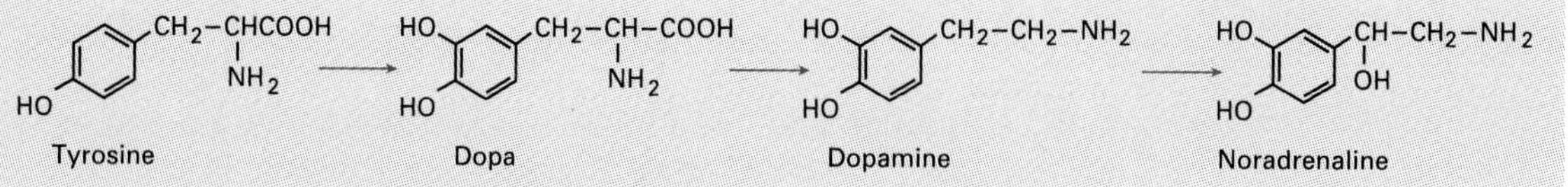

Tyrosine ($HO-C_6H_4-CH_2-CHCOOH$, NH_2) → Dopa ($(HO)_2C_6H_3-CH_2-CH(NH_2)-COOH$) → Dopamine ($(HO)_2C_6H_3-CH_2-CH_2-NH_2$) → Noradrenaline ($(HO)_2C_6H_3-CH(OH)-CH_2-NH_2$)

Figure 21.5. Pathway of biosynthesis of catecholamines from tyrosine.

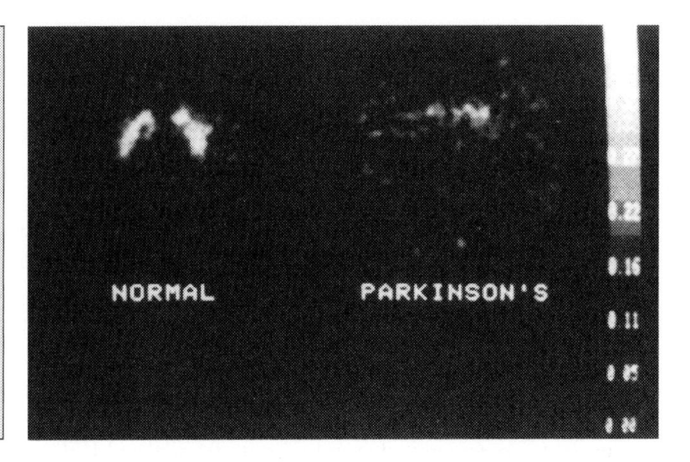

Figure 21.6. A PET scan following intravenous ^{18}F-DOPA from a normal subject (left) and a patient with Parkinson's disease (right). The scale on the right represents radioactivity in arbitrary units, increasing from black to white. Reproduced from Calne *et al.* (1985) *Nature* **317**, 246–8.

isotope in the terminals in the striatum whereas radioactivity in this brain region is markedly reduced in Randy and is similar to the levels seen in a true parkinsonian patient (Figure 21.6).

The temporary improvement in Randy's symptoms after dopamine-receptor stimulation (with bromocriptine) or administration of the dopamine precursor L-DOPA emphasises the resemblance of the drug-induced condition in the patient to PD. The highly-selective neuronal loss seen in the patient is also typical of PD and has suggested that MPTP-induced PD may serve as a useful model of the disease in experimental animals.

The entry of many compounds into the brain is restricted by the so-called blood–brain barrier and amines such as dopamine are unable to enter the brain readily; they must therefore be synthesized locally. There are, however, specific transport systems permitting the entry of amino acids into the brain, hence the use of L-DOPA which is an amino acid and the direct precursor of dopamine. The L-DOPA accumulates in the brain and is converted locally to dopamine and therefore serves as a chemical replacement therapy. DOPA is more effective than tyrosine in this process because of the involvement of tyrosine in many other pathways. Tyrosine hydroxylase is also rate-limiting in dopamine biosynthesis. The inclusion of carbidopa prevents metabolism of DOPA to dopamine by DOPA decarboxylase in liver and in peripheral nerves, minimising peripheral side effects and also allowing lower doses of L-DOPA to be administered. By directly stimulating dopamine receptors bromocriptine can also potentiate the response. L-DOPA therapy in PD is one of the best examples of rational biochemical treatment of a disease. A more recent, if controversial, treatment has been the use of foetal cell transplants into the brain to attempt to 'cure' PD.

Some years ago, a famous boxer showed the typical 'punch-drunk' symptoms of boxer's dementia, a well-characterised syndrome with symptoms similar to those of senile dementia and a matter of great concern to boxing authorities and sports medicine generally. However, people suffering chronically from PD also often develop symptoms of dementia. In order to avoid bad publicity, the boxing fraternity asserted that the boxer had PD rather than boxer's dementia, whereas the regulatory authorities concerned were of the opposite opinion! Combined PET scans with ^{18}F-DOPA and $^{15}O_2$ would easily have distinguished between the two disorders:

^{18}F-DOPA would show any deficit in dopamine in the substantia nigra in parkinsonism whereas measurement of the cerebral metabolic rates of oxygen (from using $^{15}O_2$) would have demonstrated the cortical lesions typical of dementia.

MPTP is not, in itself, toxic but is oxidised via an intermediate ($MPDP^+$) to the presumed toxic product MPP^+ by monoamine oxidase (MAO) in glial cells (astrocytes) surrounding the neurones. The normal role of MAO in the brain is, of course, to inactivate the monoamine neurotransmitters after their uptake into neuronal or glial cells. The MPP^+ is, by chance, a good substrate for the dopamine transporter molecule located specifically in the pre-synaptic membrane of dopamine neurones, hence its selective accumulation in, and toxicity to, these cells. Sensitivity of animals to the toxic effects of MPTP is remarkably variable, primates being highly vulnerable, but dogs, cats and mice are somewhat more resistant. Even goldfish are vulnerable but rats are highly resistant! The reasons for these species differences are unclear. The oxygen electrode data indicate that MPP^+ inhibits the ADP-stimulated oxidation of NAD^+-linked substrates, but not succinate oxidation, pinpointing its action at complex I of the respiratory chain, probably at or near to the site of action of rotenone. The resultant inhibition of ATP synthesis is presumed to result in cell death, although other mechanisms have been invoked such as free radical-induced actions. The chain of events in MPTP-induced toxicity is therefore probably as shown in Figure 21.7.

The causes of PD are unknown. It was once felt to be a consequence of infection especially during the enkephalitis epidemics of 1917–26. Some of those suffering from 'enkephalitis lethargica' entered into a state of suspended animation – conscious and aware yet motionless and speechless, like living 'zombies' – until partially revived by administration of L-DOPA several decades later. The dramatic effects of L-DOPA in these patients are evocatively documented in the book *Awakenings* by the British neurologist Oliver Sacks, more recently made into a popular film starring Robert de Niro and Robin Williams. The recent reports of MPTP-induced parkinsonism have tended to support the view that environmental toxins rather than infection may be a cause of the disease, especially since many compounds in the environment resemble MPTP, e.g. Paraquat and other herbicides, although no firm link has yet been established. Patients with PD can be subdivided into two groups: a minority (approximately 20%) with multiple system atrophy (MSA), a failure of the central nervous system, and the majority with pure autonomic failure (PAF), in which the peripheral nervous system is affected. These two subgroups can be distinguished by monitoring their response to the drug clonidine. Normal subjects and those with PAF respond with a large increase in blood levels of growth hormone. Distinguishing these two subgroups should allow more effective treatment since, for example, patients with the MSA variety do not respond to human foetal tissue implants.

Biogenic amines such as serotonin and the catecholamines have a role to play in the control of mood as well as movement and hence monoamine oxidase inhibitors have been used clinically in the treatment of depression. By inhibiting the enzyme they are believed to increase the availability of the neurotransmitter in the synaptic cleft. The other major class of antidepressant drugs are the so-called 'tricyclic antidepressants' which block the plasma membrane neurotransmitter transport systems, especially the serotonin transporter, again enhancing synaptic levels of monoamines. The dopamine transporter appears to be the site through which cocaine exerts its effects.

Figure 21.7. Possible mechanism of MPTP-induced neurotoxicity. Modified from Tipton and Singer (1993) *Journal of Neurochemistry* 61, 1191–1206.

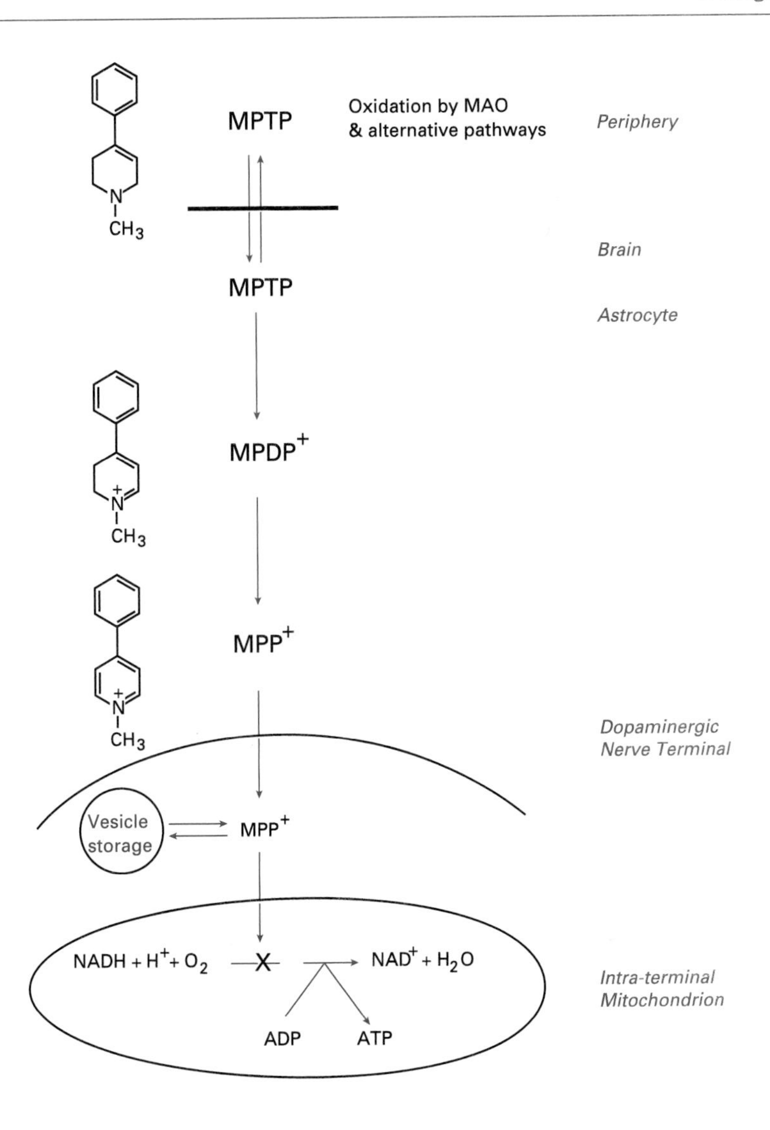

Further Questions

1. The starting point for DOPA and catecholamine synthesis is tyrosine. Is this an essential amino acid? What other biologically active molecules are produced from tyrosine? (See also Problem 19.)
2. Why might foetal cell transplants into the brain 'cure' PD? What cells might you choose?
3. What other examples of environmental toxins do you know? (See also Problems 2, 4 and 14.)

Connections

- Use this problem to review the electron transport chain and oxidative phosphorylation. What does the oxygen electrode (Figure 21.4) measure? What is meant by the P:O ratio? What is rotenone and what does it do? (See also Problem 4.)
- Find out what you can about glucose uptake and the transport molecules involved. In what respects do the glucose and dopamine transporters resemble one another? (See also Problem 13.)
- In phenylketonuria, tyrosine metabolism is also deranged. What are the neurological consequences of untreated phenylketonuria and do they resemble those of parkinsonism? (See also Problem 19.)
- There are relatively few examples of successful rational drug design for treatment of human disease. What other examples have you come across? (See also Problem 16.)
- There are a number of examples of the body converting innocuous compounds into much more toxic products, as in this problem. Do you know of any other examples? (See also Problems 2 and 14.) Can you find out why fluoroacetate and hypoglycin (from the unripe fruit of the Jamaican ackee tree) are poisonous?

References

Attwell D and Bouvier M (1992) Cloners quick on the uptake. *Current Biology* **2**, 541–3. Review on cloning and localisation of neurotransmitter transporters.

Björklund A (1993) Better cells for brain repair. *Nature* **362**, 414–15.

Changeux J-P (1993) Chemical signalling in the brain. *Scientific American* **269**(5), 58–62.

Gage F H (1993) Fetal implants put to the test. *Nature* **361**, 405–6. *News and Views* article.

Langston J W, Ballard P, Tetrud J W and Irwin I (1983) Chronic parkinsonism in humans due to a product of meperidine-analog synthesis. *Science* **219**, 979–80. Original report.

Raichle M E (1985) Progress in brain imaging. *Nature* **317**, 574–6. *News and Views* article.

Sacks O (1990) *Awakenings*, Harper Collins, New York *and* in paperback (1991), Pan Books, London.

Sedvall G (1990) PET imaging of dopamine receptors in human basal ganglia: relevance to mental illness. *Trends in Neurosciences* **13**, 302–8.

Singer T P, Trevor A J and Castagnoli N (1987) Biochemistry of the neurotoxic action of MPTP. *Trends in Biochemical Sciences* **12**, 266–70.

Snyder S H (1991) Vehicles of inactivation. *Nature* **354**, 187. *News and Views* article on structure and function of neurotransmitter transporters.

Tanner C M (1989) The role of environmental toxins in the etiology of Parkinson's disease. *Trends in Neurosciences* **12**, 49–54.

Tuomanen E (1993) Breaching the blood–brain barrier. *Scientific American* **268**(2), 56–60.

22 WHEN THE MAN FAILS, THE WOMAN TAKES OVER

Introduction

First impressions count! How often have we heard that, yet how often we can be wrong. For instance, we are all certain that we can judge whether someone whom we have just met is male or female. But would we be right?

Sex determination is a complex process which starts in the embryo and is completed at puberty. It depends first and foremost on the chromosomal (or genotypic) sex of the individual. However, the phenotypic expression of gender is a complex process and many abnormalities can arise during sexual development, such that the phenotypic gender may be at variance with the chromosomal sex.

The development of the reproductive tracts in male and female mammals is illustrated in Figure 22.1. Initially male and female embryos are indistinguishable; both possess indifferent gonads and two paired ductal systems, the Wolffian and Müllerian ducts (named after the 18th and 19th Century German embryologists, Caspar Wolff and Johannes Müller). The crucial trigger in sex determination is the production of a testis determining factor (TDF) by a gene on the Y chromosome in male embryos. TDF causes the indifferent gonads to become testes and secrete two types of hormone. Androgens stimulate the development of the male internal reproductive organs from the Wolffian ducts and also cause the external genitalia to virilise. Anti-Müllerian hormone (AMH) causes the Müllerian ducts to regress.

Since there is no Y chromosome in females, TDF is absent. Thus the gonads fail to receive the stimulus required to form testes; instead they develop into ovaries. In the absence of testes, neither androgens nor AMH are produced so the Wolffian ducts regress and the external genitalia fail to virilise, remaining female in appearance. In the absence of AMH, the Müllerian ducts develop into the female internal organs. The male is thus the determined sex while the female is the default situation.

Androgens are obviously crucial for male sex determination. However, to respond to androgens, target tissues, such as the Wolffian ducts and primordial genitalia, must possess functional androgen receptors. These intracellular proteins bind androgens specifically and act in the nucleus as regulators of gene transcription. They are part of a protein family whose members also include the receptors for other types of steroid hormones, the thyroid hormones and retinoids. This nuclear receptor superfamily has provided some of the most important clues about how gene expression is regulated.

The patients featured in this problem suffer from abnormal sexual development.

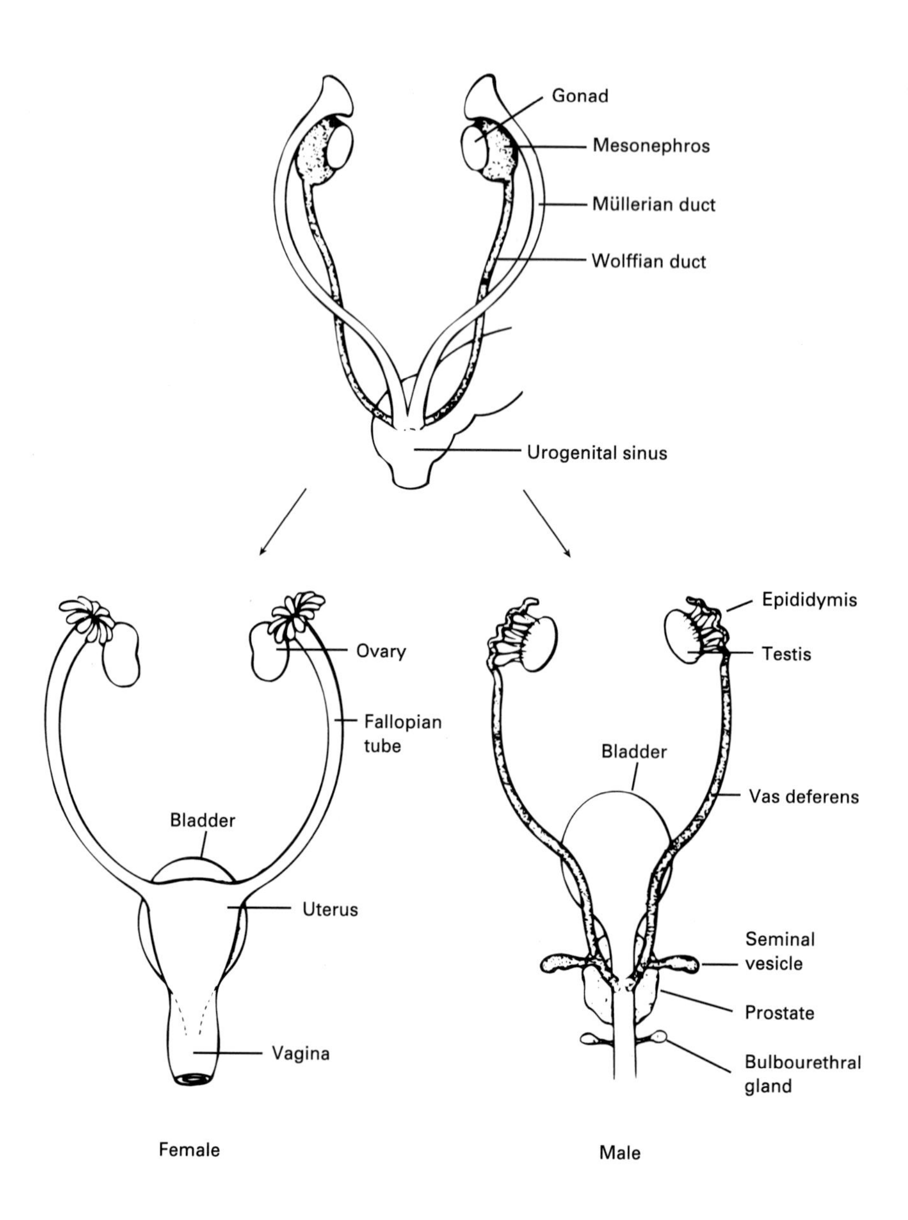

Figure 22.1. Development of the male and female reproductive systems. Adapted from Austin and Short (1982) *Reproduction in Mammals – Embryonic and Fetal Development* Cambridge University Press.

The Problem

Brenda R. and Liz W. are two unrelated post-pubertal women who were referred to a gynaecology clinic with primary amenorrhoea. In outward appearance each was unambiguously female, though their axillary and pubic hair was rather scanty. However, the patients possessed a male genotype and their plasma testosterone levels were typical of post-pubertal males. Despite this, genital skin fibroblasts, cultured from biopsies, failed to respond to normal concentrations of the active metabolite of testosterone, 5α-dihydrotestosterone (DHT).

The patients' androgen receptors (AR) were examined using these genital skin fibroblasts as a convenient source of receptors. Cell extracts were prepared and analysed for androgen receptors by denaturing polyacrylamide gel electrophoresis and immunoblotting with an antibody against normal receptor (Figure 22.2). In addition, samples of the extracts were incubated with radiolabelled DHT to examine the ability of the receptors to bind steroid (Table 22.1).

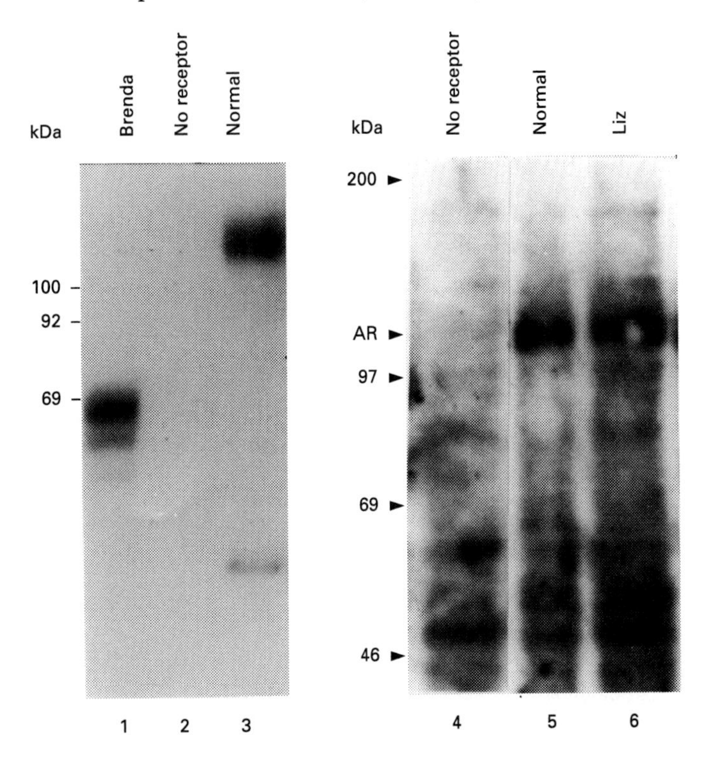

Figure 22.2. Detection of androgen receptors in extracts of skin fibroblasts. Fibroblast proteins were separated by electrophoresis in polyacrylamide gels containing sodium dodecyl sulphate (SDS–PAGE). Proteins were transferred from the gel to nitrocellulose by blotting. The blots were then incubated with radiolabelled antibody against the androgen receptor and bound antibody was detected by autoradiography. The different lanes of the gels contained extracts prepared from: Brenda R.'s cells (lane 1), cells devoid of receptors (lanes 2 and 4), cells from normal subjects (lanes 3 and 5) and Liz W.'s cells (lane 6). Adapted from Brown *et al.* (1990) *Molecular Endocrinology* 4, 1759–72; and Marcelli *et al.* (1990) *Molecular Endocrinology* 4, 1106–16 using photographs kindly supplied by Dr Terry Brown and Dr Michael McPhaul.

Table 22.1. Ability receptors to bind androgens. Extracts from skin fibroblasts were incubated with radiolabelled DHT and the apparent equilibrium dissociation constant (K_d) was calculated.

Cell extract	K_d
Normal subject	0.5×10^{-9} M
Liz W.	2.5×10^{-9} M
Brenda R.	(no binding detected)

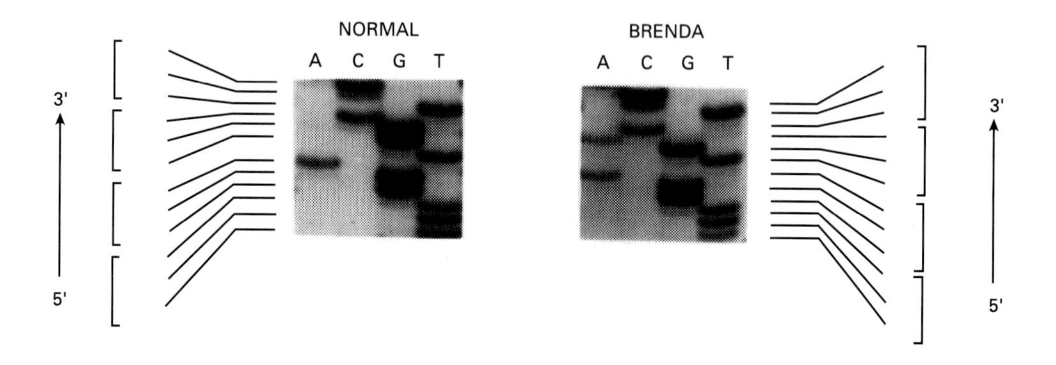

Figure 22.3. Nature of the mutation in Brenda R.'s androgen receptor gene. DNA samples from the patient's fibroblasts and from a normal subject were amplified by PCR and sequenced. The nucleotide sequence should be read from the bottom of the gel towards the top; this will correspond to the coding strand of the gene in the 5′ →3′ direction. A band in lane A indicates the presence of an adenosine nucleotide at that point in the sequence; similarly bands in lanes C, G and T correspond to cytidine, guanosine and thymidine nucleotides, respectively. Adapted from Marcelli *et al.* (1990) *Journal of Clinical Investigations* 85, 1522–8, using a photograph kindly supplied by Dr Michael McPhaul.

Next, DNA was isolated from fibroblasts of each patient. Knowledge of the nucleotide sequence of the human androgen receptor gene allowed the polymerase chain reaction (PCR) technique to be used to amplify specific regions of the receptor gene in each patient. Sequencing of these amplified DNAs can be used to reveal the mutation in each patient. In both cases the mutation was located in the same general region of the receptor gene. The sequence derived for the coding strand of this region of Liz's receptor gene is shown below with the sequence arranged in triplet codons:

Normal	5′	GAC TCC GTG CAG CCT GTA	3′
Liz	5′	GAC TCC ATG CAG CCT GTA	3′

The results for Brenda's gene are shown in Figure 22.3. The autoradiograph shows the section of the sequencing gel in which her mutation lies.

The ability of the receptors to stimulate gene transcription was assessed in a test system based on a 'reporter' gene, chloramphenicol acetyltransferase (*cat*). To enable this bacterial gene to be expressed in eukaryotic cells and used to test the functional activity of androgen receptors, it has been 'equipped' with a promoter from an androgen-responsive eukaryotic gene. The system works like this: the *cat* gene with a heterologous androgen-responsive promoter is transfected into eukaryotic tissue culture cells along with the androgen receptor gene which we wish to test. A normal receptor gene will express functional androgen receptors. These receptors will bind androgen and the receptor–steroid complexes will interact with the heterologous promoter, switching on the *cat* reporter gene. CAT enzyme activity is then easily detected and measured in cell extracts. In contrast, if the cells have been transfected with a mutant androgen receptor gene, functional receptors will not be produced, therefore the *cat* reporter gene will not be switched on and cell extracts will be devoid of CAT enzyme activity. Figure 22.4 shows the results of testing the patients' androgen receptor genes in this manner.

Questions

1. The investigation revealed that the patients had a male genotype. What test would show this?
2. What effect will the mutation in Liz W.'s androgen receptor gene have on the amino acid sequence of the receptor?
3. What is the nucleotide sequence of the region of Brenda R.'s androgen receptor gene shown in Figure 22.3? What effect will the mutation have on the amino acid sequence of her receptor? To work this out, decode the nucleotide sequence which you have obtained from the autoradiograph in triplet codons, indicated by the square brackets [ ]; the Genetic Code is in Appendix 6.
4. How do these mutations explain the experimental data in Figure 22.2?
5. Examine the data in Table 22.1; K_d provides an inverse measurement of the binding affinity of DHT for the receptor. What can you deduce about the ability of the patients' receptors to bind androgens? Can your deductions be explained by the effects of the gene mutations? Is it possible to predict which part of the receptor protein is responsible for steroid binding?
6. What do the results of the *cat* reporter assays (Figure 22.4) indicate about the functional activity of each patient's receptors?
7. Would the genital skin fibroblasts from either patient be expected to respond to supranormal concentrations of DHT?
8. Why do the patients express the female phenotype and have primary amenorrhoea?

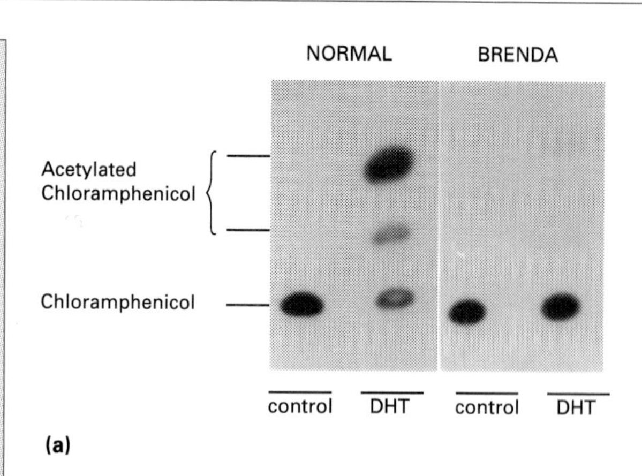

(a)

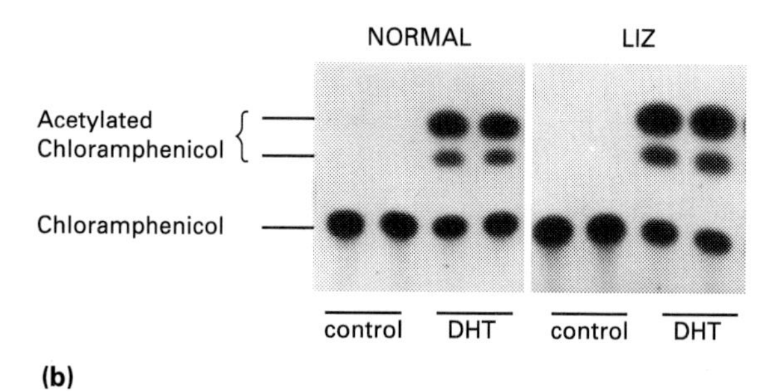

(b)

Chloramphenicol Acetyltransferase Activity

150

100

50

0

Normal

Liz

0

control

0.1

1.0

10

DHT (nM)

(c)

Figure 22.4. Functional assay of androgen receptors using a *cat* reporter gene. CAT enzyme activity was assayed by incubating cell extracts with radiolabelled chloramphenicol and then detecting acetylated chloramphenicol products by thin-layer chromatography and autoradiography. Panel A tests Brenda R.'s receptor gene; panel B tests Liz W.'s gene. Panel C shows a follow-up experiment in which the transfected cells were incubated in medium containing different concentrations of DHT; Liz's gene and the normal gene. Brenda's gene was completely inactive. Adapted from Brown *et al.* (1990) *Molecular Endocrinology* 4, 1759–72; and Marcelli *et al.* (1990) *Molecular Endocrinology* 4, 1106–16, using photographs kindly supplied by Dr Terry Brown and Dr Michael McPhaul.

Commentary

Both patients suffer from complete androgen insensitivity. Their chromosomal sex is male, which could be established from the karyotype (46 XY) of easily accessible somatic cells such as peripheral lymphocytes. To determine the karyotype, grow the cells in tissue culture in the presence of a mitogen to stimulate cell division, then add colchicine to inhibit formation of the mitotic spindle and accumulate the cells in metaphase. Lyse the cells to release the condensed metaphase chromosomes, then stain and examine them by light microscopy.

Since each patient possesses a Y chromosome, the gonads will have developed into testes under the influence of TDF. The testes are the source of the plasma androgens. However, the external genitalia and general physique are characteristically female, suggesting that the target tissues had failed to respond to the androgenic stimulus both *in utero* and post-natally.

Insensitivity to androgens prompted an examination of their androgen receptors. Figure 22.2 shows that both patients synthesize a protein which cross-reacts with the receptor antibody. However, both patients' receptors are abnormal. Brenda's is smaller than normal and is completely unable to bind androgens (Table 22.1). Liz's receptor, though of normal size, has a reduced affinity for androgens. These observations are explained by the mutations in their receptor genes. PCR provides a rapid and accurate method for determining the existence and nature of a mutation in a very small sample of DNA. The sequence of Liz's gene has a point mutation G → A. Decoding the affected codon using the Genetic Code, remembering that thymine in DNA is substituted by uracil in mRNA, shows that a valine residue has been replaced with a methionine. Figure 22.3 compares part of Brenda's gene with the normal and gives the following nucleotide sequences:

Normal	5′	TTTGGATGGCTC	3′
Brenda	5′	TTTGGATGACTC	3′

Again there has been a point mutation G → A in the protein-coding region. However, this time the change is from a tryptophan codon (TGG) into a stop signal (TGA) so Brenda's receptor protein is prematurely terminated.

Since Brenda's truncated protein cannot bind DHT (Table 22.1), the steroid binding domain must reside in the *C*-terminal region of the receptor (remember protein synthesis starts at the *N*-terminus). This is also the region in Liz's receptor where the valine has been replaced by a methionine. Although her receptor still binds DHT, it is less effective than the normal receptor (Table 22.1) suggesting that the valine residue plays a crucial role in steroid binding.

In the *cat* system, Brenda's receptor is completely unable to act as a gene activator (Figure 22.4a); this is not surprising in view of its inability to bind steroid. Liz's receptor is able to induce *cat* reporter gene expression (Figure 22.4b), but it requires considerably higher concentrations of DHT (Figure 22.4c), consistent with its lower affinity for steroid (Table 22.1). All these results suggest that genital skin fibroblasts from Liz, but not from Brenda, would respond to supranormal concentrations of androgens.

Having established that both patients are chromosomal males with functionally defective androgen receptors, their development as phenotypic females can be explained. Being unable to respond to androgens, their Wolffian ducts regressed and their external genitalia did not virilise. However, their testes did produce AMH, which caused regression of their Müllerian ducts, so they lack female internal organs. This, and the absence of ovaries, are the causes of the amenorrhoea. Undescended abdominal testes would be found on laparotomy, a procedure which would also confirm the absence of female internal organs.

The extent of androgen insensitivity and subsequent feminisation depends on the precise effect of the mutation on receptor function. Since an individual's upbringing and consequent psychological health are crucially dependent on correct gender assignment at birth, babies with ambiguous genitalia should be very carefully investigated. Many cases of testicular feminisation go undetected until amenorrhoea or infertility are investigated.

Further Questions

1. Find out about the structure of steroid receptors and their functional domains. The amino acid sequence in a centrally-located region of the receptor is highly conserved in all the members of the nuclear receptor superfamily. What functional significance has this domain? Mutations in this domain, especially those which alter the conserved amino acids, also result in non-functional receptors. In what ways would receptor function be altered and how would you investigate these? What are 'zinc fingers' and what is their relevance to transcriptional regulation?

2. Mutations abolishing receptor function are also known to occur in other members of the receptor superfamily. What symptoms would be displayed by patients with defective receptors for vitamin D or thyroid hormones?

3. Female athletes in the Olympic Games are sex-tested to screen out males who are considered to have an unfair physical advantage. From what you have learned about the patients in this case, do you think karyotyping would be the appropriate test to use?

4. Breast cancer is one of the most common cancers in women. Since the breast is oestrogen-dependent for growth, many breast tumours can be controlled by anti-oestrogen chemotherapy. Anti-oestrogens are compounds which block the binding of oestrogens to the oestrogen receptor. Unfortunately, some breast cancers are oestrogen-independent so this form of chemotherapy is unsuitable. To decide whether anti-oestrogen therapy is appropriate, the oncologist must know whether or not the tumour contains oestrogen receptors. Based on the information in this problem, what assay procedures can you devise to determine the oestrogen receptor content of a breast biopsy sample? What would be the relative merits of each assay?

5. As mentioned in this problem, steroid hormones regulate gene transcription; so also do polypeptide hormones. By what mechanisms do these two groups of hormones act – what are the similarities and differences? (See also Problems 15 and 18.)

Connections

- Review the different classes of steroid hormones, their sites of synthesis, their principal roles and how their concentrations in the blood are regulated. (See also Problem 7.)
- The steroid receptor superfamily also includes the thyroid receptors. Check that you know which the thyroid hormones are, how they act and what a goitre is.
- Find out how PCR is carried out and about its role in modern diagnostic techniques.
- Review binding assays and make sure you understand how the interaction of a hormone with its receptor can be quantified.
- This problem used an antibody to detect the androgen receptor. Review the properties of antibodies (both polyclonal and monoclonal) and the different types of assay that employ them. (See also Problem 12.)
- Fibroblasts are frequently used in screening. Do you understand why? What roles do these cells have in the body? What other types of human cell can be used in screening for genetic defects? (See also Problems 8 and 18–20.)
- There are other superfamilies of receptors. Give examples that you have come across and check that you know how they operate. (See also Problems 18 and 21.)

References

Hughes M R, Malloy P J, Kieback D G, Kesterton R A, Pike J W, Feldman D and O'Malley B W (1988) Point mutations in the human vitamin D receptor gene associated with hypocalcemic rickets. *Science* **242**, 1702–5.

Imperato-McGinley J and Canovatchel W J (1992) Complete androgen insensitivity. *Trends in Endocrinology and Metabolism* **3**, 75–81.

King R J B (1992) Effects of steroid hormones and related compounds on gene transcription. *Clinical Endocrinology* **36**, 1–14.

McLaren A (1991) The making of male mice. *Nature* **351**, 96. *News and views* article.

McPhaul M J and Marcelli M (1992) Molecular defects in the androgen receptor causing androgen resistance. *Journal of Investigative Dermatology* **98**, 97S–99S.

Mullis K B (1990) The unusual origin of the polymerase chain reaction. *Scientific American* **262**(4), 36–43.

Rhodes D and Klug A (1993) 'Zinc fingers'. *Scientific American* **268**(2), 32–9.

Usala S J and Weintraub B D (1991) Thyroid hormone resistance syndromes. *Trends in Endocrinology and Metabolism* **2**, 140–4.

Vines G (1992) Last Olympics for the sex test? *New Scientist* **135** (1828), 39–42.

Watson J D, Gilman M, Witkowski J and Zoller M (1992) In *Recombinant DNA*. Scientific American Library, New York. Chapter 6 The polymerase chain reaction, and Chapter 9 Controlling eukaryotic gene expression.

Weatherall D J (1991) *The New Genetics and Clinical Practice*. Oxford University Press, Oxford, New York and Tokyo. Several useful chapters.

Appendix 1 BLOOD CONSTITUENTS (REFERENCE RANGES)

Values are for adults and refer to plasma/serum except as indicated*. Further data will be found in *New England Journal of Medicine* **314**, 39–49 (1986).

Electrolytes and blood gases

Osmolality	280–298 mmol/kg
pCO_2 (arterial)	4.5–6.1 kPa
pO_2 (arterial)	11–15 kPa
pH (arterial)	7.35–7.45
Sodium	134–147 mM
Potassium	3.5–5.0 mM
Chloride	96–106 mM
Bicarbonate	21–30 mM
Phosphate	0.7–1.4 mM
Sulphate	0.25–0.38 mM
Calcium	2.1–2.7 mM
Magnesium	0.7–1.3 mM
Copper	12–30 μM
Zinc	12–24 μM
Lead	<2.4 μM

Haemoglobin and blood cells

Haemoglobin*	13–17 g/dl (males)
	11–15 g/dl (females)
Erythrocytes (red cell count)*	4.5–6.5 × 10^{12}/l (males)
	3.9–5.6 × 10^{12}/l (females)
Packed cell volume (PCV)*	0.41–0.53 l/l† (males)
	0.35–0.48 l/l† (females)
Mean corpuscular volume (MCV)*	76–96 fl
Mean corpuscular haemoglobin (MCH)*	26–33 pg/cell
Mean corpuscular haemoglobin concentration (MCHC)*	31–35 g/dl
Leukocytes (white cell count)*	4–6 × 10^9/l
Platelets*	2–4 × 10^{11}/l
Reticulocytes*	0.5–2.5% red cells
Glucose 6-phosphate dehydrogenase (erythrocyte)*	5–15 U/g Hb
Pyruvate kinase (erythrocyte)*	13–17 U/g Hb
Iron	10–30 μM
Total iron-binding capacity (TIBC)	40–70 μM

Nitrogenous constituents and amino acids	Ammonia (NH_4^+)	12–60 μM
	Urea	2.5–7.0 mM
	Creatinine	53–124 μM
	Uric acid	0.18–0.42 mM
	Alanine	0.2–0.3 mM
	Glutamine	0.45–0.75 mM
	Citrulline	10–30 μM
	Phenylalanine	0–0.12 mM
Organic constituents	Glucose (fasting)	3.5–5.5 mM
	Lactate	0.6–1.8 mM
	Pyruvate	0.03–0.11 mM
	Citrate	0.08–0.16 mM
	Bilirubin (total)	3–15 μM
Vitamins	Ascorbic acid (C)	20–100 μM
	Biotin	2.5–5.5 μM
	Cholecalciferol (D_3)	0.07 μM
	Cobalamin (B_{12})	0.15–0.67 nM
	Folic acid	15–80 nM
	Niacin (nicotinic acid)	50 μM
	Pantothenic acid	1.3 μM
	Retinol (A)	0.3–2.1 μM
	Riboflavin (B_2)	0.07–0.1 μM
	Thiamine (B_1)	0.2–0.3 μM
	α-Tocopherol (E)	20–45 μM
Lipids (fasting)	Triglyceride	0.6–3.2 mM
	Cholesterol	3.7–6.8 mM
	Fatty acids (total)	0.2–0.8 mM
Proteins	Total protein	60–84 g/l
	Albumin	35–50 g/l
	Globulins (total)	23–35 g/l
	IgG	7–15 g/l
	Lipoprotein (HDL)	3–8 g/l
	Ceruloplasmin	0.2–0.4 g/l
Clotting factors	Fibrinogen	2–4 g/l
	Plasminogen	0.3 g/l
	Prothrombin	0.1 g/l

Enzymes*	Acid phosphatase	4–11
	Alkaline phosphatase	100–300
	Aldolase	2–12
	Creatine kinase (CK)	20–150
	Lactate dehydrogenase (LDH)	45–90
	Amylase	70–300
	γ-Glutamyl transferase (GGT)[††]	7–45
	Aspartate transaminase (AST)[§]	10–40
	Alanine transaminase (ALT)[§§]	5–35
	Serum lipase	28–280
	Cholinesterase	39–51
	α-Fucosidase	4.5–14.0
	α-Galactosidase	0.22–0.50
	β-Galactosidase	0.4–1.1
	α-Mannosidase	0.35–0.90
	N-Acetyl-α-glucosaminidase	0.15–0.40
	N-Acetyl-β-glucosaminidase	12–25
	p-Nitrocatechol sulphatase	0.04–0.12
	α-Glucosidase	0.1–0.25

* Enzyme activities in units/l (U/l)
† PCV also expressed as % = l/l × 100.
†† Serum γ-glutamyl transpeptidase (γ-GT)
§ Serum glutamate-oxaloacetate transaminase (SGOT)
§§ Serum glutamate-pyruvate transaminase (SGPT)

Appendix 2 URINARY CONSTITUENTS (REFERENCE RANGES)

The daily (24 hour) excretion of many urinary constituents will vary considerably from subject to subject depending on diet, age, sex, exercise, etc. Therefore, the values listed in this Appendix are only very approximate. Further data will be found in *New England Journal of Medicine* **314**, 39–49 (1986).

	Volume	1–2 litres/day
	pH	4.8–7.5 (mean 6.0)
Nitrogen compounds	Total nitrogen	600–1400 mmol/day
	Urea	300–640 mmol/day
	Ammonia (NH_4^+)	60–110 mmol/day
	Uric acid	2–6 mmol/day
	Creatinine	7–17 mmol/day
	Amino acids	7–21 mmol/day
	Protein	0–150 mg/day
Electrolytes	Titratable acidity (H^+)	15–50 mmol/day
	Sodium	180–440 mmol/day
	Potassium	40–80 mmol/day
	Calcium	5–12 mmol/day
	Magnesium	4–8 mmol/day
	Chloride	100–260 mmol/day
	Sulphate (as S)	18–56 mmol/day
	Phosphate (as P)	23–52 mmol/day
	Lead	0–500 nmol/day
	Copper	0–1.6 μmol/day
Porphyrins	δ-Aminolævulinic acid	0–40 μmol/day
	Porphobilinogen	0–16 μmol/day
	Coproporphyrin	0–380 nmol/day
	Urobilinogen	0–7 μmol/day
	Uroporphyrin	0–49 nmol/day
Others	Glucose	negative
	Ketones	negative
	Citrate	1.0–5.7 mmol/day
	17-ketosteroids	3–56 μmol/day
	17-hydroxysteroids	8–22 μmol/day

Appendix 3 CONCENTRATIONS OF HORMONES IN HUMAN PLASMA

The concentrations of some hormones are often expressed in activity units. Where possible these have been converted into molar concentrations. The ranges quoted are approximate and may be affected by many factors such as age, sex, race, environment, posture, diurnal or menstrual cycles, nutritional state, feeding, drugs taken and exercise. In addition, some hormones are present mainly in a form bound to a variety of specific and non-specific carrier proteins; the concentration of the free active form of a hormone may be much less than that indicated in the Appendix.

Peptide/protein hormones	Adrenocorticotropin (ACTH)	$2–20 \times 10^{-12}$ M
	Calcitonin	$4–9 \times 10^{-12}$ M
	Erythropoietin	<50 mg/l
	Follicle stimulating hormone (FSH)	5–150 U/l (women) 3–18 U/l (men)
	Glucagon	$20–50 \times 10^{-12}$ M
	Growth hormone	$50–400 \times 10^{-12}$ M
	Insulin	$20–100 \times 10^{-12}$ M (fasting)
	Luteinising hormone (LH)	2–180 U/l (women) 3–20 U/l (men)
	Parathyroid hormone (PTH)	$3–65 \times 10^{-12}$ M
	Prolactin	$0.08–6 \times 10^{-9}$ M
	Thyroid stimulating hormone (TSH)	0.3–4.0 mU/l
	Vasopressin (antidiuretic hormone)	$0.5–10 \times 10^{-12}$ M
Steroid hormones	Aldosterone	$50–250 \times 10^{-12}$ M
	Cortisol	$150–600 \times 10^{-9}$ M (peak)
	Oestradiol	$0.08–1.3 \times 10^{-9}$ M (women) $< 0.18 \times 10^{-9}$ M (men)
	Progesterone	$20–100 \times 10^{-9}$ M (post-ovulation) $0.5–2.0 \times 10^{-9}$ M (pre-ovulation)
	Testosterone	$5–40 \times 10^{-9}$ M (men) $0.5–3.2 \times 10^{-9}$ M (women)
	1,25-dihydroxy vitamin D_3	$60–150 \times 10^{-12}$ M
Catecholamines	Adrenaline	$0.1–0.2 \times 10^{-9}$ M
	Noradrenaline	$1–2 \times 10^{-9}$ M
Thyroid hormones	Triiodothyronine (T_3)	$1–3 \times 10^{-9}$ M (total)
	Thyroxine (T_4)	$60–150 \times 10^{-9}$ M (total)

Appendix 4 INCIDENCE OF GENETIC AND CHROMOSOME DISORDERS

Monogenic Disorders

Representative examples are given for monogenic disorders where the gene defect has been identified. Incidences in the general population refer to the European/US white population and should be regarded as very approximate in most cases. Further data will be found in Weatherall D J (1991) The frequency and clinical spectrum of genetic diseases, Chapter 2, pp 4–38. In *The New Genetics and Clinical Practice* (3rd edn), Oxford University Press, Oxford.

(a) inborn errors of metabolism

Disorder	Affected gene product	Incidence in general population (per 1000 live births)	Incidence in subpopulations (per 1000 live births)
Acute intermittent porphyria	uroporphyrinogen synthase	0.01	3.0 Afrikaner
Argininosuccinase deficiency	argininosuccinase	0.014	
Congenital adrenal hyperplasia	steroid 21-hydroxylase	0.1–0.2	2.0 Eskimo
Cystathioninaemia	homoserine dehydratase	0.009	
Galactosaemia	galactose 1-phosphate uridyl transferase	0.008–0.02	
Gaucher's disease	glucocerebrosidase	0.013	0.4 Ashkenazi Jews
Glucose 6-phosphate dehydrogenase deficiency	glucose 6-phosphate dehydrogenase (X-linked)	0.05 (boys)	50 US Blacks
Glycogen storage disorders	various	0.017 (total)	
Histidinaemia	histidine deaminase	0.06	
Homocystinuria	cystathionine synthetase	0.01	
Lesch–Nyhan syndrome	hypoxanthine-guanine phosphoribosyl transferase (X-linked)	0.003 (boys)	
Maple syrup urine disease	branched-chain ketoacid decarboxylase	0.005	6.0 Mennonite
Methylmalonic acidaemia	methylmalonyl-CoA mutase	0.017	
Mucopolysaccharidoses	various	0.03 (total)	
Phenylketonuria (PKU)	phenylalanine hydroxylase	0.06–0.1	0.2–0.5 Celtic
Propionic acidaemia	propionyl-CoA carboxylase	0.003	
Tay–Sachs disease	*N*-acetyl hexosaminidase A	0.004	0.3 Ashkenazi Jews

(b) **other monogenic disorders**

Disorder	Affected gene product	Incidence in general population (per 1000 live births)	Incidence in subpopulations (per 1000 live births)
α_1-Antitrypsin deficiency	α_1-anti trypsin	0.1–0.5	
Cystic fibrosis	cystic fibrosis transmembrane conductance regulator	0.5	0.01 Afro-Oriental
Duchenne muscular dystrophy	dystrophin (X-linked)	0.2–0.3 (boys)	
Haemophilia A	clotting factor VIII (X-linked)	0.1 (boys)	
Hypercholesterolaemia	LDL receptor	2.0	
Marfan syndrome	fibrillin	0.04	
Osteogenesis imperfecta	procollagen	0.04–0.1	
Sickle cell disease	β-globin	0.1	10–20 Afro-Caribbean
β-Thalassaemia	β-globin	0.05	5–10 Mediterranean

Chromosome Disorders

Syndrome/Disorder	Incidence in population (per 1000 live births)	Incidence in subpopulations (per 1000 live births)
Turner's (45, X)	0.1–0.2	
Klinefelter's (47, XXY)	1–2	
Down's (trisomy 21)	1–2	30–40 if mother aged 45+
Edward's (trisomy 18)	0.1–0.2	
Fragile X	0.9 (boys)	

Appendix 5

AMINO ACID DESIGNATIONS

Amino acids may be given one- or three-letter abbreviations according to the table below.

Amino acid	Three-letter abbreviation	One-letter abbreviation
Alanine	Ala	A
Arginine	Arg	R
Asparagine	Asn	N
Aspartic acid	Asp	D
Cysteine	Cys	C
Glutamic acid	Glu	E
Glutamine	Gln	Q
Glycine	Gly	G
Histidine	His	H
Isoleucine	Ile	I
Leucine	Leu	L
Lysine	Lys	K
Methionine	Met	M
Phenylalanine	Phe	F
Proline	Pro	P
Serine	Ser	S
Threonine	Thr	T
Tryptophan	Trp	W
Tyrosine	Tyr	Y
Valine	Val	V

Appendix 6

THE GENETIC CODE

The Code is read in triplet codons. The base corresponding to the first position (5′) is given along the left side of the table, the second (middle) base is along the top and the third (3′) base is on the right. Thus, the codon -5′ AUG 3′- in the mRNA corresponds to methionine (Met). The codons UAA, UAG and UGA are termination (stop) codons. See Appendix 5 for the amino acid three-letter abbreviations.

1st Position	2nd Position: U		C		A		G		3rd Position
U	UUU	Phe	UCU	Ser	UAU	Tyr	UGU	Cys	U
	UUC	Phe	UCC	Ser	UAC	Tyr	UGC	Cys	C
	UUA	Leu	UCA	Ser	UAA	Stop	UGA	Stop	A
	UUG	Leu	UCG	Ser	UAG	Stop	UGG	Trp	G
C	CUU	Leu	CCU	Pro	CAU	His	CGU	Arg	U
	CUC	Leu	CCC	Pro	CAC	His	CGC	Arg	C
	CUA	Leu	CCA	Pro	CAA	Gln	CGA	Arg	A
	CUG	Leu	CCG	Pro	CAG	Gln	CGG	Arg	G
A	AUU	Ile	ACU	Thr	AAU	Asn	AGU	Ser	U
	AUC	Ile	ACC	Thr	AAC	Asn	AGC	Ser	C
	AUA	Ile	ACA	Thr	AAA	Lys	AGA	Arg	A
	AUG	Met	ACG	Thr	AAG	Lys	AGG	Arg	G
G	GUU	Val	GCU	Ala	GAU	Asp	GGU	Gly	U
	GUC	Val	GCC	Ala	GAC	Asp	GGC	Gly	C
	GUA	Val	GCA	Ala	GAA	Glu	GGA	Gly	A
	GUG	Val	GCG	Ala	GAG	Glu	GGG	Gly	G

GLOSSARY OF MEDICAL AND SCIENTIFIC TERMS

This Glossary defines medical, scientific and genetical terms used in the Problems. It assumes a familiarity with the meanings of routinely-used biochemical terms which may be found in any good text book of biochemistry. Where part of a definition is set in **bold** type, there is a separate glossary entry.

Acanthosis — thickening of the inner layer of cells of the epidermis by abnormal increase in the number of cells. *Acanthosis nigricans*, dark warty growth on the skin, especially in skin folds.

Acidaemia — abnormally high blood acidity (low pH), see **acidosis**.

Acidosis — abnormally high acidity of body tissues and fluids because of failure of the mechanisms regulating the acid–base balance.

Acini — singular, *acinus*; bunch of small sacs or cavities lined with secreting cells in a gland (Greek, 'bunch of grapes').

Acute — disease of rapid onset with severe symptoms and brief duration (contrast with **chronic**).

-aemia — suffix pertaining to the blood.

Aetiology — (USA, *etiology*) cause of a disease.

Agonist — substance that triggers a response in a cell, such as a hormone or neurotransmitter (opposite, **antagonist**).

Akinesia — loss of muscular responsiveness, arrest of motion.

Allele — one of a number of alternative forms of a gene that occupy a given genetic locus.

Amenorrhoea — absence of menstrual periods.

Aminoaciduria — presence of abnormal amounts of amino acids in the urine.

Amniocentesis — withdrawal of **amniotic** fluid which surrounds the foetus by piercing the amniotic sac through the abdominal wall. Used to obtain foetal cells.

Amniotic — concerned with the *amnion*, the membrane that initially forms over the dorsal part of the embryo but expands to cover it completely within the *amniotic cavity*.

Anabolism — building up; in metabolic terms, making more complicated molecules by biosynthesis (opposite, **catabolism**).

Anaemia — reduction in the quantity of haemoglobin in the blood.

Androgen — steroid hormone (e.g. testosterone) that stimulates the development of the male sex characteristics.

Aneurysm — balloon-like swelling in the wall of an artery weakened by degenerative disease.

Angiogram — X-ray picture produced following the injection of an X-ray opaque dye into an artery.

Anisocytosis — excessive variation in size between individual **erythrocytes**.

Anoxaemia — less than the normal amount of oxygen in the blood (see **anoxia**).

Anoxia — condition where body tissues receive inadequate amounts of oxygen.

Antagonist — inhibitor of hormone or neurotransmitter action (opposite, **agonist**).

Antibiotic	any of a very diverse group of organic compounds usually produced by microorganisms that selectively inhibit or kill other microorganisms (e.g. penicillin).
Antihypertensive	drug used to treat raised blood pressure (**hypertension**).
Apolipoprotein	protein moiety of a lipoprotein that transports lipids in the blood.
Apoprotein	protein part of any complex or conjugated protein.
Arachnodactyly	having abnormally long and slender fingers, reminiscent of spider.
Arteriole	small artery.
Asymptomatic	not showing any signs of disease whether disease is present or not.
Ataxia	unsteady gait resulting from brain's failure to regulate posture and the strength and direction of limb movements.
Atheroma	degeneration of the walls of arteries due to the formation of fatty plaques or scar tissue which limits blood circulation and predisposes to thrombosis. *Atherosclerosis* is a disease in which *atheromatous* plaques develop.
Atrophy	wasting away of a tissue or organ.
Autoimmune	of diseases suspected of being caused by inflammation and deterioration of the body's tissues due to attack by its own antibodies.
Autopsy	dissection and examination of a body after death (e.g. to discover cause of death).
Autoradiograph	pattern formed on photographic film when radioactive material (e.g. in the spots on a chromatogram) is placed against it.
Autosome	any chromosome that is not a **sex chromosome**.
Benzodiazepines	group of drugs that act as tranquillisers and hypnotics.
Bilirubin	pigment found in blood derived from haem breakdown (see **jaundice**).
Bilirubinaemia	an excess of **bilirubin** in the blood.
Bioassay	determination of the biological activity of a compound such as a hormone.
Biopsy	removal of a small piece of tissue from a patient for examination by histology or other analysis. Usually carried out with a hollow needle.
Blood–brain barrier	structural and physiological barriers which prevent the movement of most blood components into brain tissue or cerebrospinal fluid or vice versa.
Blot	transfer of nucleic acids or proteins from a gel to a nitrocellulose or nylon sheet prior to analysis (see **Northern**, **Southern** and **Western blots**).
Calcification	deposition of calcium salts, e.g. in the formation of bone.
Canaliculi	small channels (singular, *canaliculus*), e.g. in compact bone.
Carcinogen	substance or agency that produces cancer, e.g. chemicals in cigarette smoke, ionising radiation (see **precarcinogen**).
Carcinoma	cancer or tumour that arose in an **epithelium**.
Cardiomegaly	abnormal enlargement of the heart.
Cardiovascular	(*Cardiovascular system*) the heart together with the networks of blood vessels (pulmonary and systemic).
Catabolism	breaking down of food or storage materials, e.g. to provide energy (opposite, **anabolism**).
Cataract	opacity of the lens of the eye causing blurred vision.
cDNA	copy DNA or complementary DNA produced from messenger RNA by using reverse transcriptase (RNA-dependent DNA polymerase).
Cerebellum	largest part of the hind brain, responsible for maintaining muscle tone and synchronising the activity in groups of muscles under involuntary control.
Chemotherapy	treatment of disease with chemical compounds (drugs); compare radiotherapy.

Chorionic villus	part of the *chorion*, the embryonic membrane that totally surrounds the embryo.
Chorionic villus sampling	removing a small piece of thin tissue through the cervix for examination to detect possible abnormalities in the foetus (see **prenatal diagnosis, amniocentesis**).
Chromosome spread	a picture of the complete set of chromosomes prepared by treating cells in certain ways enabling the chromosomes to be identified (see **karyotype**).
Chronic	of long duration (of a disease) involving slow changes, possibly of gradual onset (opposite, **acute**). Does not imply anything about the severity of the disease.
Cirrhosis	condition in which the liver responds to some insult (e.g. excess alcohol intake) by producing interlacing strands of fibrous tissue resulting in a characteristic knobbly appearance.
Clathrin	fibrous protein that lines 'coated pits' at the cell surface. Arranged in a characteristic polyhedral fashion. Involved in **endocytosis**.
Coma	state of unrousable unconsciousness; the patient is said to be *comatose*.
Congenital	of a condition that is present since birth, e.g. *congenital malformation*.
Cornea	the transparent circular part of the front of the eyeball.
Corpuscle	a blood cell.
Cystathioninuria	presence of *cystathionine* in the urine.
Cystic fibrosis	hereditary disease in which a thick mucus is produced that obstructs intestinal glands, pancreas and bronchi. Severe respiratory infections are a common complication.
Cytokines	substances, usually peptides or proteins, that act as growth factors, i.e. affect the growth, division or **differentiation** of cells (e.g. interferon, interleukins, platelet-derived growth factor).
Cytosol	'watery' material remaining when all the organelles and membranes have been removed from an homogenate of cells by centrifugation.
Dementia	chronic or persistent disorder of mental processes due to brain disease, including memory disorders, changes in personality, disorientation etc. (e.g. Alzheimer's disease).
De novo	'from new'; synthesis of compounds during metabolism 'from scratch', i.e. from simple precursors.
Dermatology	study and treatment of skin diseases.
Detoxification	process by which the body 'makes safe' or detoxifies foreign compounds that may be deleterious (see **xenobiotic**).
Diabetes mellitus	disorder in which glucose is not taken up and metabolised by cells due to lack of, or insensitivity to, the pancreatic hormone insulin.
Diagnosis	process of determining the nature of a disorder by an evaluation of the patient's signs and symptoms, medical history, laboratory tests, etc. *Differential diagnosis*, distinguishing a condition whose signs and symptoms are shared by other conditions.
Differentiation	increasing specialisation of cells during development.
Diploid	of organisms whose cells (except for the gametes) have two (duplicate) sets of chromosomes.
Dissemination	wide distribution in an organ of the body, e.g. with reference to disease organisms or pathological changes.
Distal	situated away from the origin or point of reference, e.g. that part of a limb furthest away from the body (opposite, **proximal**).
Diurnal	occurring every day; *diurnal rhythm*, metabolic or behavioural rhythm with a cycle of 24 hours.

Dolichostenomelia having long thin limbs.

Domain structurally or functionally defined section of a protein.

Dominant (genetics) property possessed by some **alleles** of solely determining the **phenotype** when present in either the **homozygous** state or as one member of a **heterozygous** pair when they mask the effects of the other allele (the **recessive** allele).

Down's syndrome form of mental subnormality resulting from a chromosomal defect, the presence of three copies (instead of two) of chromosome 21 (trisomy 21); formerly 'mongolism'.

Dysplasia abnormal development of a tissue.

Effector something that brings about activity in a muscle or gland, e.g. a nerve impulse or hormone.

Encephalitis inflammation of the brain.

Endemic (of diseases) occurring regularly in a particular region or population.

Endocrine of hormones, the secretion of the ductless glands.

Endocytosis process by which cells take up material by an invagination of the plasma membrane to form vesicles enclosing external material (opposite, **exocytosis**).

Endogenous originating within the body or cell (opposite, **exogenous**).

Endothelial pertaining to the *endothelium*, a squamous **epithelium** that lines internal surfaces such as heart, blood vessels and fluid-filled cavities and glands.

Endotoxin bacterial toxin that remains within the cell that produces it (contrast **exotoxin**).

Enterotoxin bacterial toxin that affects the intestine.

Enzymopathy disease resulting from the (genetic) failure to produce an active enzyme.

Epidemiology study of *epidemics*; the occurrence/spread of diseases within populations with a view to finding a means of control or future prevention.

Epilepsy any of a group of disorders of brain function producing recurrent seizures of sudden onset.

Epithelium plural, *epithelia*; a sheet of cells lining any external or internal surface, e.g. epidermis, mucous membranes.

Erythema abnormal flushing of the skin caused by dilation of the capillaries.

Erythrocyte mature red blood cell (**corpuscle**) lacking a nucleus (see also **reticulocyte**).

Erythropoiesis process of red blood cell production in the bone marrow.

Exocrine glands whose secretion is drained by ducts (contrast **endocrine**).

Exocytosis process by which molecules, etc., are secreted from cells. Membrane-bound vesicles fuse with the plasma membrane to release their contents to the exterior (opposite, **endocytosis**).

Exogenous originating outside the body or cell (opposite, **endogenous**)

Exon in a gene, a sequence of DNA which is transcribed to form part of the mature mRNA which is translated into protein in the cytoplasm. Separated from the next exon in the gene by an **intron**, or non-coding sequence, which does not form part of the mature mRNA.

Exotoxin a toxin secreted by a bacterium (contrast, **endotoxin**).

Extracellular matrix the macromolecular ground substance between the individual cells in a connective tissue. Composed of collagen and proteoglycans, secreted by **fibroblasts**.

Familial of a condition found in some families but not others (often inherited).

Fibroblast widely distributed cell type characteristic of connective tissue (see **extracellular matrix**).

Fibrosis thickening and scarring of connective tissue usually due to inflammation or injury.

Gastrectomy	surgical removal of all (*total gastrectomy*) or part (*partial gastrectomy*) of the stomach.
Gastric (gastro)	pertaining to the stomach.
-gen (-genic)	suffix meaning 'giving rise to'.
Genome	the genetic complement of an organism or of a single cell; the total number of genes.
Genotype	genetic constitution of an organism (see **phenotype**).
Gestation	period during which the fertilised egg develops into a baby that is ready to be born.
Glucagonoma	glucagon-secreting tumour of the α-cells of the **islets of Langerhans** of the pancreas.
Glucogenic	literally 'glucose generating', e.g. of some amino acids; others are **ketogenic**.
Glucosuria	glucose in the urine.
Glycogenosis	glycogen storage diseases in which abnormally large amounts of glycogen are deposited in cells, especially liver and/or muscle.
Gonadotropin	(gonadotrophin) any of a number of pituitary hormones that act on the gonads (testis or ovary).
Gout	disease associated with the precipitation of crystals of uric acid in the tissues and especially the joints, due to **hyperuricaemia**. Local reaction produces inflammation and intense pain.
Habitus	an individual's general physical appearance.
Haemagglutinin	substance that causes the clumping together (agglutination) of red blood cells.
Haematology	study of blood and its formation.
Haemocytometer	a glass chamber of known volume on a microscope slide used for visual counting of the various types of blood cells.
Haemodialysis	technique for removing toxic waste products from the blood by dialysis, usually performed on patients with kidney failure.
Haemoglobinopathy	disease or condition resulting from the production of an abnormal, or insufficient, haemoglobin (see **sickle cell anaemia, thalassaemias**).
Haemolysis	destruction of red blood cells (*haemolytic streptococci*, bacteria that cause *haemolysis*).
Haemostasis	arrest of bleeding involving coagulation and contraction of blood vessels. Also applied to surgical procedures used to stop bleeding.
Hepatic	of the liver. *Hepatic portal system*, blood vessels draining from the gut to the liver.
Hepatitis	inflammation of the liver with a variety of causes, e.g. viruses, toxins or immunological abnormalities.
Hepatobiliary	of the liver passages for bile formation.
Hepatocyte	principal cell type of the liver.
Hepatoma	malignant tumour of the liver.
Hepatomegaly	abnormal enlargement of the liver.
Heterozygous	of a *heterozygote*, a **diploid** organism that has inherited different **alleles** of any particular gene from each parent (see **homozygous**).
Hirsute	hairy. *Hirsutism*, presence of coarse, dark hair on face, chest, back or abdomen in females.
Histology	study of the detailed structure of tissue by means of special staining techniques and light or electron microscopy.
HLA complex	*human leukocyte antigen*; a series of gene families that code for proteins expressed on the surface of most nucleated cells. Involved in tissue recognition and graft rejections. If two individuals have the same HLA types they are histocompatible and can accept each other's tissue grafts.
Homeostasis	maintenance of a stable internal environment.
Homocystinuria	having homocystine in the urine.

Homozygous	of a *homozygote*, a **diploid** organism that has inherited the same **allele** of a particular gene from both parents (see **heterozygous**).
Hybridisation	technique for determining if two DNA molecules have complementary sequences by reassociating their single strands and determining the extent of double helix formation. Can also be done with DNA–RNA.
Hyper-	prefix meaning 'raised'.
Hyperammonaemia	raised level of ammonia in the blood.
Hyperbilirubinaemia	raised level of **bilirubin** in the blood.
Hypercholesterolaemia	abnormally high blood cholesterol concentration.
Hyperextensible	unusually stretchable, used of joints and skin in collagen diseases.
Hyperglycaemia	raised level of glucose (sugar) in the blood.
Hyperkalaemia	abnormally high blood potassium concentration.
Hyperlacticacidaemia	raised level of lactic acid (lactate) in the blood.
Hypernatraemia	abnormally high blood sodium concentration.
Hyperoxaluria	raised level of oxalic acid in the urine.
Hyperplasia	excessive development of a tissue or organ due to an increase in the number of cells; abnormal increase in cell proliferation.
Hypertension	raised blood pressure.
Hyperuricaemia	raised level of uric acid (urate) in the blood.
Hypo-	prefix meaning 'lowered'.
Hypocalcaemia	abnormally low blood calcium concentration.
Hypochromic	less coloured, pale.
Hypoglycaemia	decreased level of glucose in the blood.
Hypokalaemia	abnormally low blood potassium concentration.
Hyponatraemia	abnormally low blood sodium concentration.
Hypoparathyroidism	condition due to decreased production or activity of parathyroid hormone.
Hypothyroid	subnormal activity of the thyroid gland.
Hypotonic	of a solution having a lower osmotic pressure than another, usually less than physiological (contrast *isotonic, hypertonic*).
Hypoxia	a deficiency of oxygen in the tissue.
Immuno-inhibition	use of an antibody to inhibit an enzyme activity.
Inborn error	(of metabolism) a genetic defect resulting in a metabolic or other deficiency.
Infarction	death of part or all of an organ that occurs when the artery bringing in blood is obstructed by a blood clot (thrombosis).
In situ	literally, 'in place'. Usually used to describe the method of detecting a specific DNA sequence in a tissue section by **hybridisation** with a labelled DNA **probe**.
Insulinoma	insulin-producing tumour of the β-cells of the **islets of Langerhans** of the pancreas.
Intoxication	symptoms of poisoning due to ingestion of toxic material (e.g. heavy metals, alcohol).
Intravenous	into a vein.
Intron	in a gene, a sequence of DNA between two **exons**. Intron sequences are transcribed by RNA polymerase to form part of the primary RNA transcript of the gene, but are then spliced out so that they do not form part of the mature mRNA which is translated in the cytoplasm to form protein. In many genes introns may make up 80%–90% of the nucleotide sequence. A few genes, such as some of those for interferons, do not possess introns.
In utero	(of a foetus) 'in the uterus'.

In vitro	'in glass', i.e. in the test tube, in the laboratory.
In vivo	'in life', i.e. in a living organism.
Ischaemia	an inadequate flow of blood to part of the body caused by constriction or blockage. *Myocardial ischaemia* affects the **myocardium**.
Islets of Langerhans	small clumps of endocrine cells scattered throughout the pancreas. The α cells produce glucagon, the β cells produce insulin.
Isoenzyme(s)	(also *isozyme*) physically distinct forms of an enzyme catalysing the same reaction.
Jaundice	a yellow coloration of the skin and whites of the eyes indicating **hyperbilirubinaemia**.
Karyotype	a representation of the complete chromosomal complement of a cell in which individual mitotic chromosomes are arranged in pairs in order of size (see **chromosome spread**).
Ketoacidosis	**acidosis** resulting from the accumulation of ketone bodies (acetoacetate, D-β-hydroxybutyrate, acetone) in the blood.
Ketogenic	(of an amino acid) giving rise to ketone bodies (see **ketoacidosis, ketosis**).
Ketosis	poisoning caused by an accumulation of ketones (acetoacetate, D-β-hydroxybutyrate, acetone) in the blood (see **ketoacidosis**).
Kwashiorkor	severe malnutrition widespread in the tropics that sets in after weaning.
Laparotomy	a surgical incision into the abdominal cavity.
Laryngeal	pertaining to the larynx.
Lectin	a member of a group of (mostly) plant proteins that can agglutinate animal cells ***in vitro*** by binding to specific carbohydrates on the cell surface. Some act as **mitogens**.
Lesion	a region of damaged tissue with impaired function resulting from damage due to disease or injury. By analogy, a defect in a metabolic pathway.
Leukaemias	a group of **malignant** diseases in which the bone marrow produces increased numbers of certain types of **leukocytes**. May be **acute** or **chronic**. Overproduction of immature or abnormal leukocytes suppresses the production of other blood cells and leads to increased susceptibility to infection, **anaemia**, bleeding, etc.
Leukocyte	a white blood cell or **corpuscle**.
Ligand	usually a small molecule that binds to a large one such as a protein or a nucleic acid.
Lipophilic	'fat loving', e.g. of dyes or histological stains that stain fatty areas.
Lipoprotein	protein–lipid complex of which a number of types are found in the blood, e.g. HDL (high-density lipoprotein), LDL (low-density lipoprotein); based on their behaviour in the **ultracentrifuge**.
Lymphocytes	a class of **leukocyte** also present in lymph nodes, spleen, thymus and bone marrow; involved in immunity.
Lymphoid	of tissue responsible for production of **lymphocytes** or antibodies.
Macro-	prefix meaning 'large'.
Malignant	denoting a tumour that invades and destroys tissue and can spread (contrast, *benign*); also applied to any other life-threatening disorder, e.g. *malignant hypertension*.
Marasmus	severe wasting disease of infants (**chronic** starvation).

Megaloblast abnormal form of any of the cells that are precursors of **erythrocytes**, usually large (as the prefix 'megalo' implies). Megaloblastic **anaemias** are characteristic of vitamin B_{12} or folic acid deficiency.

Meiosis nuclear division resulting in the daughter nuclei each containing half the number of chromosomes of the parent (i.e. haploid rather than **diploid**), in the production of gametes (sperm or ova).

Menopause cessation of menstruation (contrast, *menarche*, the start of menstruation).

Mesothelioma tumour of the pleura, peritoneum or pericardium, often associated with exposure to asbestos fibres.

Metacarpal of the metacarpals, bones of the hand.

Metaphase stage in **meiosis** or **mitosis** when the chromosomes have become aligned around the equator of the cell prior to cell division.

Metastasis distant spread of a **malignant** tumour from its site of origin (**primary** tumour) via the blood or lymph or through body cavities such as the peritoneum. Highly malignant tumours have a greater tendency to *metastasise* to form **secondary** tumours.

Metatarsal of the metatarsals, bones of the foot.

MHC (*major histocompatibility complex*), a cluster of genes encoding the major antigens involved in tissue recognition and graft rejection. In humans it is known as the **HLA complex**.

Micro- prefix meaning 'small'.

Microcytosis presence of abnormally small **erythrocytes**; a feature of certain **anaemias** and **haemoglobinopathies**.

Mitogen substance that promotes **mitosis** and (usually) cell division.

Mitosis process of nuclear division when each member of a duplicated chromosome segregates into a daughter nucleus. The daughter nuclei now contain identical sets of **diploid** chromosomes; cell division follows to produce two **diploid** cells genetically identical to the parent cell (see also **meiosis**).

Monoclonal of antibodies derived from a single clone of **lymphocytes** and which are then all identical (contrast **polyclonal**).

Monocyte type of **leukocyte** having a kidney-shaped nucleus: functions to ingest foreign particles and bacteria (abbreviated from *mononuclear leukocytes*).

Monogenic condition controlled by a single gene (contrast, **polygenic**).

Mononucleosis condition where the blood contains an abnormally high number of **monocytes**. *Infectious mononucleosis* is commonly referred to as glandular fever, an infectious disease associated with Epstein-Barr virus.

Morbidity state of being diseased.

Morphology form and structure of the body as distinct from its physiology, etc.

Mortality (rate) the incidence of death in a population in a given period.

Mucolipidosis disease characterised by intracellular accumulation of complex glycolipids (mucolipids).

Mutagen agent (chemical, radiation) that can cause a **mutation**.

Mutation change in the sequence information of DNA which may result in an alteration in the **phenotype** of an organism. Those in the germ line cells will be inherited; those in other body cells (**somatic** cells) will not (see also **point mutation**).

Myocardium middle of the three layers that form the wall of the heart, forming the greater part of the heart muscle. *Myocardial infarction* – death of a segment of heart muscle following interruption of the blood supply (see **ischaemia**).

Myoglobinuria presence of myoglobin in the urine.

Nasopharyngeal of the nasopharynx, that part of the pharynx or throat continuous with the posterior openings of the nasal passages.

Natriuretic an agent that promotes the excretion of sodium salts in the urine.

Necrosis death of some or all of the cells of an organ or tissue.

Neonatal new born, within the first four weeks of life.

Neural tube in embryology, the structure from which the brain and spinal cord develop. Failure to form properly results in *neural tube defects* (e.g. spina bifida).

Neurotoxic poisonous to nerve cells.

Neutrophil a type of **leukocyte** characterised by the possession of a lobed nucleus and fine granules in the cytoplasm (a granulocyte). Capable of ingesting and killing bacteria.

Nitrogen balance equilibrium state of the body in which nitrogen intake (in the form of amino acids) and excretion (as urea, etc.) are equal.

Nodule small swelling.

Northern blot technique in molecular biology for identification of specific types (sequences) of RNA following electrophoresis in a gel. Carried out by **hydridisation** with a radioactive **probe**. Named by comparison with a **Southern blot**. (See also **Western blot**).

Obesity condition where excess fat has accumulated in the body, especially **subcutaneous** fat.

Oedema (USA, *edema*), accumulation of fluid in the body tissue ('dropsy'), may be local or general. In air spaces of lung, *pulmonary oedema.*

Oesophagus (oesophageal) (USA, *esophagus, esophageal*) tube joining the throat with the stomach.

Olfaction process of being able to smell.

Oligonucleotide short sequence of nucleotides, small piece of DNA. *Oligonucleotide probe*, piece of DNA, often radiolabelled, used as a **probe** in a **Northern** or **Southern blot**.

Oligosaccharide small carbohydrate polymer.

Omental of the omentum, a double layer of peritoneum attached to the stomach, linking it with other organs such as the liver, spleen and intestines.

Oncogenes genes that can cause cancer; many produce proteins such as growth factors that normally regulate cell division but escape control in the process of **oncogenesis**.

Oncogenesis development of an abnormal growth or tumour (benign or **malignant**) (Greek, *oncos* – 'a lump').

Osteoblast cell type responsible for the formation of bone.

Osteoclast large multinucleated cell that re-absorbs calcified bone, e.g. in the process of bone remodelling.

Osteodystrophy generalised bone disease resulting from a metabolic disorder.

Osteogenesis 'making bone'. *Osteogenesis imperfecta,* impaired bone formation.

Osteoporosis loss of bone mineral resulting in brittle bones. May be localised due to infection or injury, or generalised as occurs in the elderly.

Palpate to examine the body by careful feeling with the hands and fingertips.

Parathyroid hormone hormone from the parathyroid gland that controls the distribution of calcium and phosphate in the body.

Parkinsonism having Parkinson's disease, a neurological disorder of middle-aged or elderly people characterised by tremor, rigidity and a paucity of spontaneous movements.

Patency (patent) being open, e.g. of a blood vessel.

Pathogenesis origin of a disease; *pathogen*, an agent (e.g. microorganism) causing disease.

Pathology study of disease processes by observing and analysing samples of blood, urine and faeces as well as diseased tissue obtained by **biopsy** or at **autopsy**.

PCR see **polymerase chain reaction**.

PCV packed cell volume (of **erythrocytes**, after centrifugation).

Pedigree analysis in genetics, analysis of the ancestry of a group of related individuals.

Pericellular surrounding a cell.

Peripheral near the surface or extremities. *Peripheral blood film*, blood taken from a superficial site and spread or smeared on a slide for microscopic examination.

Peritoneum serous membrane of the abdominal cavity.

Pernicious of diseases likely to result in death if untreated. *Pernicious anaemia*, an **anaemia** resulting from a deficiency of vitamin B_{12} due to lack of proper absorption.

Pertussis whooping cough, an acute contagious disease, mostly of children, caused by the microorganism *Bordetella pertussis*.

Phenotype visible or otherwise measurable physical and biochemical characters of a person resulting from the interaction of the **genotype** with the environment.

Phenylketonuria literally a condition in which aromatic keto compounds are present in the urine. Specifically an **inborn error** of metabolism in which an enzyme dealing with phenylalanine is deficient leading to urinary excretion of phenyllactate and phenylpyruvate. Causes mental deficiency when untreated.

Phototherapy treatment of disease by the use of light.

Plasma straw-coloured fluid in which the blood cells are suspended.

Poikilocytosis presence of abnormally shaped **erythrocytes**.

Point mutation **mutation** resulting from the change of a single base in DNA.

Polyclonal of antibodies derived from a number of clones of **lymphocytes** simultaneously responding to the same antigen (contrast, **monoclonal**). The normal immunological response is polyclonal.

Polydipsia abnormally intense thirst, symptom of **diabetes mellitus**.

Polygenic condition involving the collective effects of several different genes (contrast **monogenic**).

Polymerase chain reaction (PCR) a molecular biology technique that enables a specific sequence of DNA in a mixture to be amplified enormously.

Polymorph a **neutrophil**, a type of **leukocyte**.

Polymorphism condition in which a chromosome or a genetic character occurs in more than one form resulting in the existence of more than one morphological type in a population.

Polyuria producing abnormally large amounts of urine, characteristic of **diabetes mellitus**.

Porphyrias group of inherited disorders of the biosynthesis of haem.

Portal concerned with that part of the circulation draining blood from the intestines to the liver, hence *portal vein, portal circulation, portal* ***hypertension***.

Postmenopausal occurring after the **menopause**.

Post mortem 'after death', a medical examination after death (see **autopsy**).

Postnatal 'after birth'.

Postpubertal occurring after puberty.

Precarcinogen substance which may be metabolised to produce an active **carcinogen**.

Premenopausal occurring before the **menopause**.

Prenatal 'before birth'. *Prenatal diagnosis*, detecting disease in infants prior to birth (see **amniocentesis, chorionic villus sampling**).

Primary	the initial cause of a disease, site of original tumour before **metastasis** (contrast **secondary**).
Probe	usually a radioactively-labelled nucleotide or DNA sequence used to detect the presence of a particular gene sequence by **hybridisation** in a **Southern** or **Northern blot**. May also be applied to a labelled antibody used to detect specific proteins in a **Western blot**.
Proteinuria	presence of protein in the urine.
Proximal	situated close to the origin or point of reference (contrast **distal**).
Pseudohypo-parathyroidism	condition whose symptoms resemble those due to decreased secretion of parathyroid hormone.
Pulmonary	of the lungs.
Radiography	examination of the body using X-rays.
Recessive	property of **alleles** whose effects are not reflected in the **phenotype** when present as one member of a **heterozygous** pair; they only determine the phenotype when present in the **homozygous** state (opposite, **dominant**).
Recombinant	e.g. *recombinant protein*, a protein produced by genetic engineering technology, usually by inserting the appropriate gene into a microorganism that produces (expresses) the corresponding protein. 'Humulin' is human recombinant insulin, produced by this process, and used to treat **diabetes mellitus**.
Renal	pertaining to the kidneys.
Resection	e.g. *bowel resection*, taking out a piece of intestine and joining the remaining cut ends together.
Restriction enzymes	a series of microbial enzymes that specifically cleave DNA. Used in molecular biology and genetic engineering to manipulate DNA.
Restriction map	genetic map produced by using **restriction enzymes**.
Reticulocyte	immature, precursor of an **erythrocyte** named for the reticular (network) structures revealed by **histology**.
Rickets	disease of children due to lack of vitamin D; the bones do not harden properly and they bend out of shape.
Rigor mortis	stiffening of the body that occurs within about 8 hours of death due to changes in muscle tissue. Starts to disappear after about 24 hours.
Schizophrenia	severe mental disorders in which delusion and hallucinations occur. The patient usually feels that thoughts, sensations, actions are controlled or shared by others.
Sclerae	(sing. sclera). White fibrous outer layer of the eyeball.
Secondary	dependent or following on from another (**primary**) event. *Secondary metabolites*, products of (normally) minor off-shoots of a metabolic pathway. *Secondary tumour*, tumours (secondaries) arising by **metastasis**.
Sensory	relating to the information carried by the nervous system to the brain and spinal cord. Such impulses are carried by *sensory nerves*.
Serosal	basal aspect of an **epithelium** facing the blood vessels (contrast *luminal*).
Serum	**plasma** from which the proteins of the clotting system have been removed by allowing the blood to clot.
Sex chromosomes	the X and Y chromosomes which determine the sex of an individual; male XY, female XX (contrast **autosomes**).
Sickle cell anaemia	**haemoglobinopathy** resulting from a **point mutation** in the β-globin gene that produces a modified haemoglobin. The **erythrocytes** in such individuals deform

	('sickle') especially at low pO_2 and in the small capillaries, resulting in **haemolysis** and **anaemia**.
Sinusoid	small blood vessel found in certain organs such as liver and the adrenal glands.
Somatic	(of a **mutation**) relating to body cells rather than generative (germ) cells, and which will not therefore be passed on to the offspring.
Southern blot	a molecular biology technique for identifying specific DNA sequences by **hybridisation** with a (radioactively) labelled, single stranded DNA **probe**. Named after E M Southern (see also **Northern** and **Western blots**).
Splenomegaly	enlargement of the spleen.
Steatosis	infiltration of liver cells (**hepatocytes**) with fat.
Stenosis	abnormal narrowing of a passage or opening such as a blood vessel or heart valve.
Steroidogenesis	metabolic process whereby steroids are synthesized.
Striatum	region of the brain forming part of the basal ganglia and involved in the initiation and planning of movements. The dopamine pathways terminating in the striatum degenerate in **Parkinsonism**.
Subcutaneous	beneath the skin.
Symport	membrane protein that simultaneously cotransports two different solutes across a membrane in the same direction (opposite, *antiport*).
Synaptic cleft	narrow fluid-filled gap between two nerve cells.
Syndrome	group of symptoms characteristic of a particular disease.
Tetanus	condition of a muscle undergoing a continuous series of contractions due to electrical stimulation. Hence a disease characterised by paralysis, caused by the toxin produced by *Clostridium tetani* infection.
Thalassaemias	group of inherited **anaemias** in which there are a wide variety of defects in haemoglobin production (see **haemoglobinopathy**).
Thrombolysis	dissolution of a blood clot (*thrombus*) by an enzyme.
Topology	surface features.
Transduction	carrying a signal (e.g. hormonal) across the cell membrane to activate intracellular biochemical pathways.
Transfection	genetic modification of cultured cells by the introduction of DNA into the cells. The DNA may be stably incorporated into the **genome**.
Transformation	genetic modification of a bacterium by DNA which enters the cells. Also used of eukaryotic cells following infection with tumour viruses or treatment with a **carcinogen**, which gives the cells the ability to divide indefinitely.
Translocation	movement to another place; specifically the transfer of one section of a chromosome to another, thus changing the order of genes which may lead to a serious genetic disorder.
Tumorigenesis	generation of a **tumour** (see **oncogenesis**).
Tumour	swelling, usually as a result of abnormal cell proliferation (see **oncogenesis**).
Tumour suppressor gene	gene involved in the suppression of uncontrolled cell growth and division. Cancer may develop because of the loss or inactivation of these genes in **somatic** cells (see also **oncogenes**).
Ultracentrifuge	centrifuge that spins very rapidly allowing not only cell particles but also macromolecules to be sedimented or separated.
Ultrastructure	the minute structure of cells at the electron-microscopic level.

Ureotelic excreting excess nitrogen as urea, rather than as ammonia in fish (ammoniotelic) or uric acid in birds, reptiles (uricotelic).

-uria suffix pertaining to urine.

Vaccination process of giving a **vaccine** to immunise the patient against an infectious disease.

Vaccine a preparation of microorganisms or their toxins or a virus that can lead to the production of protective immunity but that does not itself cause disease.

Varices varicose veins.

Vasoactive affecting the diameter of blood vessels, especially arteries, and hence influencing blood pressure.

Vasodilation increasing the diameter of blood vessels, especially arteries, which produces a lowering of the blood pressure (opposite, *vasoconstriction*).

Venule small vein.

Virilisation result of excessive **androgen** production in females, produces a male body form, deepening of the voice.

Western blot detection of proteins following their electrophoresis in gels. The gels are placed in contact with a nitrocellulose or nylon membrane and the protein is transferred out of the gel by electrophoresis. Once on the membrane the proteins stick firmly and may be identified with a (labelled) specific antibody **probe**. Named by comparison with **Southern** and **Northern blots**.

Xanthomata yellowish swellings or nodules in the skin resulting from deposits of fat (singular, *xanthoma*).

Xenobiotic foreign (substance) to a living organism; applied to drugs, etc.

Xeroderma pigmentosum rare, inherited skin disease in which the skin is light-sensitive and there is a predisposition to skin **tumours** caused by a defect in a DNA repair enzyme.

Zymogen inactive precursor of an enzyme, the active form being produced by a specific proteolytic cleavage.

Zymogram pattern produced when enzyme proteins are subjected to electrophoresis and the electrophoretogram is treated subsequently with a substrate of the enzyme that produces a coloured product.

GUIDE TO TOPICS COVERED

Problem	Topic areas	Techniques featured
1	muscle exercise, energy metabolism, ischaemia	metabolite levels in biopsy samples
2	starvation, nutrition, protein-energy metabolism, toxin metabolism, liver damage	
3	cardiovascular disease, muscle damage, differential diagnosis	serum isoenzyme levels, electrophoresis of isoenzymes (lactate dehydrogenase, creatine kinase)
4	poisoning, pesticides, respiratory chain and oxidative phosphorylation	interpretation of oxygen electrode data
5	neonatal development, amino acid catabolism, nitrogen metabolism, dietary therapy, inborn errors	interpretation of blood metabolite levels, design of dietary therapy
6	haematology and toxicology, iron and haem metabolism, nutrition	interpretation of blood films and bone marrow histology, quantitative handling of haematological data
7	endocrinology, steroid biosynthesis, salt and water homeostasis, genetic defects, genetic screening, hormone replacement therapy	interpretation of blood metabolite levels, deduction of metabolic block, karyotype analysis
8	bone formation, protein structure, collagen structure and biosynthesis, post-translational modification of proteins, genetic mutation, genetic screening	electrophoresis (SDS-PAGE), interpretation of mutational effects, design of gene probes
9	cell biology, protein targeting, acid hydrolases, mucopolysaccharide turnover, post-translational modification of proteins	electron microscopy, pulse-labelling, evaluation of blood enzyme levels

Problem	Topic areas	Techniques featured
10	inborn errors, metabolic block, vitamin B_{12}, acid–base balance, ketoacidosis, anaemias, dietary therapy	interpretation of blood metabolite levels, effects of vitamin therapy
11	molecular biology, haematology, haemoglobin biosynthesis, epidemiology	interpretation of blood films, electrophoresis of haemoglobins, tryptic maps, identification of mutations, interpretation of gene and protein sequences
12	endocrinology, control of blood glucose, actions of glucagon and insulin, radioimmunoassay	assay and interpretation of blood hormone levels, quantitative handling of radioimmunoassay data
13	cholera, oral rehydration therapy, transport of sugar, electrolytes and water, cell receptors and signalling, toxins	
14	tissue-specific actions of carcinogens, mutagenesis, xenobiotic metabolism, detoxification	interpretation of tissue incubations
15	endocrinology, calcium homeostasis, vitamin D and parathyroid hormone, transmembrane signalling, genetic defects	interpretation of blood hormone and metabolite levels
16	hypertension, cardiovascular disease, bioactive peptides, pharmacology, rational drug design, enzyme inhibitors as pharmacological agents	interpretation of amino acid sequence data, enzyme kinetics, interpretation of enzyme inhibitor effects
17	alcohol metabolism, liver damage, liver function tests, cirrhosis	histology, interpretation of serum enzyme levels, enzyme kinetics
18	diabetes, insulin action, control of blood glucose, genetic mutations, insulin receptor structure and action	glucose tolerance test, electrophoresis (SDS-PAGE), trypsin digestion of proteins, DNA sequencing, deduction of mutational effects
19	amino acid metabolism, inborn errors, essential amino acids, brain development, prenatal diagnosis	interpretation of Western and Northern blots, pedigree analysis, use of allele-specific probes

Problem	Topic areas	Techniques featured
20	cardiovascular disease, atherosclerosis, epidemiology, plasma lipids and lipoproteins, enzyme inhibitors as therapeutic agents	plasma lipid analysis, lipoprotein electrophoresis, Southern blot, gene structure analysis, restriction enzymes
21	neurobiology, drug abuse, neurotransmitters, synaptic transmission	positron emission tomography (PET) scans, interpretation of oxygen electrode data
22	endocrinology, molecular biology, androgens and androgen receptors, hormone–receptor binding, development of the reproductive system, sex determination	hormone binding data, DNA sequence interpretation, deduction of mutations, gene activity analysis using CAT reporter gene, Western blots